BIOETHICS - MEDICAL, ETHICAL AND LEGAL PERSPECTIVES

Edited by **Peter A. Clark**

Bioethics - Medical, Ethical and Legal Perspectives

http://dx.doi.org/10.5772/62798
Edited by Peter A. Clark

Contributors

Eleanor Milligan, Kusal K. Das, M.S. Biradar, Frieder Keller, Makiko Kondo, José-Antonio Santos, Richard Howard, Danielle Cornell, Joseph Raho, Federico Nicoli, Mario Picozzi, Omar Ferrario, Ger Wackers, Lydia Kapiriri, Robert Ancuceanu, Ilora Gillian Finlay of Llandaff

Published by InTech

Janeza Trdine 9, 51000 Rijeka, Croatia

Publishing Process Manager Iva Simcic

Technical Editor SPi Global

Cover InTech Design team

First published January, 2017

Printed in Croatia

Additional hard copies can be obtained from orders@intechopen.com

Bioethics - Medical, Ethical and Legal Perspectives, Edited by Peter A. Clark

p. cm.

Print ISBN 978-953-51-2847-2

Online ISBN 978-953-51-2848-9

For EU product safety concerns:
IN TECH d.o.o., Prolaz Marije Krucifikse Kozulić 3, 51000 Rijeka, Croatia,
info@intechopen.com or visit our website at intechopen.com.

Meet the editor

Dr. Peter A. Clark, S.J. is currently Professor of Medical Ethics and Director of the Institute of Clinical Bioethics at Saint Joseph's University in Philadelphia, Pennsylvania. He is the Bioethicist for 12 acute care hospitals in Pennsylvania, Maryland and New Jersey and is responsible for the Ethics Core Curriculum of over 300 medical interns/residents in the areas of Internal Medicine, Surgery, Radiology, Emergency Medicine, Family Practice, OB/GYN, Podiatry, etc. His extensive pursuit of research in the areas of end-of-life issues, neonatology, medical education, organ transplantation and public health has resulted in more than 100 peer reviewed articles and book chapters. Dr. Clark is also the author of three books: *To Treat Or Not To Treat: The Ethical Methodology of Richard A. McCormick, S.J. As Applied To Treatment Decisions For Handicapped Newborns; Death With Dignity: Ethical and Practical Considerations For Caaregivers Of The terminally Ill and Contemporary Issues in Bioethics.* Dr. Clark is also active in numerous local, national and international professional organizations in the area of Bioethics.

Contents

Preface

The field of Bioethics is a relatively new discipline, but in reality, it is becoming more complex by the day. As science advances, more and more ethical questions are being exposed and few in the scientific field seem interested in taking the time to discern whether we should do all that can be done. Just because something now can be accomplished does that mean we should move forward and allow it to advance? Bioethicists are the ones raising the "red flags" about issues that are a concern to them and to society as a whole. These issues are complex and the complexity involves areas of medicine, science, law and ethics. Many complex questions are confronting us professionally and as human person. Should we move forward with Preimplantation Genetic Diagnosis (PGD)? Should we create children from three persons when the mitochondria from the biological mother is defective? Is medical rationing inevitable? Can organs for transplantation be created using animal and human stem cells? When does personhood begin? Should therapeutic medical procedures be allowed to develop into enhancement procedures? When do medical treatments become physiologically and qualitatively futile? Has autonomy become absolute in medicine today? How honest and reliable is medical research today? How safe are genetically modified foods? When is a new pharmaceutical discovery too expensive for society as a whole? Should we use gene editing to protect babies from disease? Should we implant chips in the brain to improve people's ability to think? Should Crispr be used to alter the genes of babies? Should we transfuse synthetic blood that would enhance performance by increasing speed, strength and endurance? These questions and more are being debated by scientists, lawyers, politicians and bioethicists around the world but the fear is that the scientists will succeed in advancing these new procedures and technologies because biotechnology corporations see a profit to be made by implementing them.

A recent study in 2015 by the Pew Research Center found that Americans had a profound distrust of scientists, a suspicion about claims of progress and a real discomfort with the ideas of meddling with human abilities. [1] There seems to be a real distrust of science and corporations in regards to science-related topics. Many fear the end result. Will the future hold a new dividing point: the enhanced versus the unenhanced? Who will be able to afford these new procedures? The fear is these new techniques will cause a greater divide in the world between the "haves" and the "have nots." This divide already exists and we cannot afford to have the divide widen. Instead, we must work to close the divide by creating technologies and medical procedures that benefit all humanity.

The safety of these new procedures and techniques need to be examined in depth. Long-term implications and consequences need to be considered because the future of humanity

1) Gina Kolata, "We Don't Trust Scientists To Make Us Better," New York Times July 27, 2016, A-20. http://www.pewreserach.org/science2015.

hangs in the balance. The long-term health effects of these procedures need to be examined and the implications of how these new procedures will impact on the quality of the environment must be considered. The public is beginning to realize that science is impacting on humanity—climate change, childhood vaccinations, genetically modified foods and animals, responsible research with nonhuman primates, etc. The issue is that there must be a group of responsible individuals who are willing to be in the forefront of these scientific gains in order to question, challenge and debate the scientists to make sure that the future of humanity is protected and will be treated with dignity and respect. It is the role of the Bioethicist today to engage in a constructive dialogue with his/her colleagues in medicine, scientific research and law so that new procedures and techniques are completely vetted from all vantage points to preserve and defend the very dignity and respect of the human person. This job may appear to be daunting at first glance, but unless those trained in philosophical and theological ethics take the lead, the future of humanity will become compromised. This book is an attempt to encourage a dialogue with our colleagues in medicine, research and law to examine these critical bioethical issues from all sides so that the best interest of patients, families and society as a whole will be protected in the future.

Dr. Peter A. Clark
Director-Institute of Clinical Bioethics
Professor of Medical Ethics
Saint Joseph's University
Philadelphia, Pennsylvania, USA

Chapter 1

Pharmacy Ethics and the Spirit of Capitalism: A Review of the Literature

Robert Ancuceanu and Ioana-Laura Bogdan

Additional information is available at the end of the chapter

http://dx.doi.org/10.5772/65128

Abstract

This chapter explores the issue of the conflict (real or potential) between the ethical imperatives that should guide the pharmacist in the typical practicing of the profession (i.e. within a pharmacy) and the economic constraints derived from the business dimension of the pharmacy. Marrying service and business in a single profession, pharmacy is supposed to balance harmoniously its two sides, if not to subject business demands to the higher societal, ethical requirements. However, such a balancing exercise is rather like dancing on a rope, and ethics may be trumped by economics, a phenomenon deplored sometimes by pharmacy academics or hospital pharmacists, and by a part of community pharmacists as well. Economics may prevail over ethics in rough forms such as selling health risk products (as it was in the past for tobacco or alcohol) or in more elusive ones, such as longer work hours and shorter counselling times, promoting or dispensing needless or ineffective products (food supplements, cosmetics, etc.), silently refusing to provide or recommend lower cost generics, etc. Ethical research in the field of pharmacy has generally been scarce, and numerous knowledge gaps remain to be filled by future investigations.

Keywords: pharmacy ethics, capitalism, professional altruism, commercialism, economic constraints, ethical breach

1. Introduction

With rare exceptions, the issue of the conflict (real or potential) between the ethical imperatives that should guide the pharmacist in the typical practicing of the profession (i.e. within a pharmacy) and the economic constraints derived from the economical dimension of the pharmacy has been almost completely ignored in the recent scientific literature, although in the past, they have been approached to some extent in the Anglo-American literature. The "role tension" or "role ambiguity," as suggestively has been described the conflict between the equally important demands of "professional altruism" and "commercialism" [1], has only faintly been reflected in the scientific publications of the past decades.

When in his seminal paper published in 1957, Ernest Greenwood synthesized and defined the five attributes of a profession, the first characteristic recognized was the authority rooted in an extensive professional education, but he added that "The professional must not use his position of authority to exploit the client for purposes of personal gratification" [2]. Not very different was the view adopted by T. H. Marshall (1963), who viewed the essence of professionalism in a single practitioner with particular abilities and individual responsibility ("which cannot be shifted onto the shoulders of others"), an individual who "is not concerned with self-interest, but with the welfare of the client" [3].

In theory, professions enjoy an extended degree of autonomy and self-regulation, and in exchange they are assumed to place the community needs above their own interests. In fact, there is widespread perception of a disjuncture between theory and factual reality: self-interest is often perceived as being satisfied ahead of the societal needs [4]. Broad studies across professions however, using rigorous methodologies, are lacking and many questions have only been partially answered if at all. Is this situation the same for all professions or are there differences among various occupations? Is this situation the same in different geographic regions and cultures, or are there differences among different countries and traditions? It is known that ethical decisions are influenced by cultural and value differences [5], and despite the levelling tendencies of the globalization, it is doubtful that in the field of pharmacy, ethical expectations and actions are now uniform across borders. The potential ethical conflict between the professional side and business side of pharmacy has been repeatedly asserted in the ethical literature, but its ways of expression, extent and facilitating variables seem to have been rarely investigated in depth. In this chapter, we will explore how this potential conflict between societal needs and economic self-interest is reflected in the ethical literature, based on theoretical considerations or empirical research.

2. Pharmacy: the double face of Janus

The dual character of pharmacy has been perceived and discussed in the scientific and professional literature for more than 100 years, at the beginning of the twentieth century being argued that an absolute divorce between the two sides (commercial and professional) was impossible [6]. Pharmacy has been described as presenting "a unique combination of professional and commercial elements" [7]. In 1943, A. Weinlein in an unpublished MA thesis on the pharmacy as a profession in Wisconsin made similar statements on the dual nature of pharmacy [8].

Pharmacy is a service profession, but it also has a business dimension, especially due to the fact that particular health goods (medicines, cosmetics, medical devices, etc.) are dispensed to patients and money change hands in this process. Business is traditionally seen as apt for those focused on financial gains, whereas service professions are seen as apt for those valuing altruism and service to others. Moreover, business and service professions are seen as competitive and in a direct conflict [9]. Authorities themselves have looked at pharmacists as professionals in some contexts while "reducing them to a mercantile level" in others [10]. The opposition between the two has been concentrated by E.C. Hughes (1958) in two short Latin phrases: *caveat emptor* (let the buyer—by extension, consumer—beware) and *credat emptor* (let the buyer trust) [9].

The eclectic nature of pharmacy (as "half business, half profession") has been rejected at least at times by pharmacy academics that have increasingly seen themselves and their graduates as "militant upholders of complete professionalization" [11]. The active lexical transitioning from "retail pharmacy" to "community pharmacy" has been viewed by others as the expression of the profession's own awareness of a strain between the business side and the healthcare side of pharmacy [12].

Dessing and Flameling have argued that universal ethical principles apply to both business persons and professional practitioners (care providers) [13] (and thus the distinction between the two sides would be irrelevant), but their proposition is more "what ought be" rather than "what is," and there are data indicating that a discrepancy between values and facts may in fact exist, despite the fact that a business ethics has been recognized and developed in the past decades [14].

3. Is pharmacy a quasi-profession?

The business dimension of pharmacy has often led to questioning its sociological status as a profession, being sometimes relegated to that of a *quasi-profession, incomplete* or a *marginal* one [15]. The profession status has been won based on the social role of the pharmacist in procurement, preparation and assessment of drugs, a role which has gradually vanished with the rise of the modern pharmaceutical industry and the occurrence of the premixed and prepackaged medicines [4, 16, 17]. This change has been described as a loss of the compounding function that relegated the pharmacy to the status of "another link in the chain of distribution" for the new healthcare industry [18].

In the new context, the community pharmacy setting (the "retail pharmacy") has been said to encapsulate "the most non-professional aspects of the profession" (unlike hospital pharmacy, for instance) [15].

As a first argument, the ignoring of the "no advertising" rule has been invoked in the United States where it rarely seems to be applied [15]. This rule, already expressed by Greenwood, refers to the fact that professions discourage advertising [2], because allowing it would undermine the professional authority, entrusting the client with critical abilities in selecting competing professionals, as if the client would be able to judge the quality of a professional service [15]. As a second argument, the subordination of the professional goals to personal ones (noncommitment to altruistic values and goals) was cited by Denzin and Mettlin as a fact

[15], although the contrary vision is usually held by the profession (in line with the ethics codes adopted by it [17]). A third argument advanced for the marginality of pharmacy profession is the acceptance by pharmacists to sell (non-professional items and objects) [15], pursuing a profit rather than perceiving a fee. Additional arguments against a full professional status of pharmacy (community pharmacy in particular) have been derived from the absence of a body of knowledge to be gained by socialization and the absence of a unified organization controlling its members, presumably because of the large heterogeneity of the profession manifested by an increased number of subspecializations [15].

The arguments of Denzin and Mettlin against the incomplete status of the pharmacy profession have been in 1995 refuted by Dingwall and Wilson, who argued that applying them to other professions such as medicine or law (and for some arguments even most other professions), the results would be largely similar [19].

In the United States, in the past, a prevalence of the business spirit over the professional role of pharmacists has been implied from the lack of involvement in purely professional associations [11]. In the majority, if not all European countries, though, this aspect is irrelevant, because according to law, practicing pharmacists have to belong to a professional organization [20].

The "noble profession of medicine," often perceived as one of the most disinterested and altruistic of human occupations, one ignoring mercantile considerations with a quasi-Olympian serenity, has been shown to be in a situation not very different from pharmacy with respect to the conflict between its institutional role and economic pressures, professionalism "giving way to entrepreneurialism" [21]. And a century earlier, Shrady deplored "the growth of commercialism in medicine" [22]. But even in 1922, Fischelis argued that physicians do bookkeeping, although they do not call it "commercial medicine" or lawyers in their practice do have business administration activities, without getting a "commercial" label [23].

There are authors who deny the existence of a sharp distinction between (health) professions and businesses, arguing that there is ethics in modern business on the one hand and that professionals are not wholly altruistic, on the other [24]. Furthermore, it may be argued (as one owner pharmacist from the United Kingdom affirmed) that an ethical decision may often be in the same time a good commercial decision in the long run, even when looked in isolation the pharmacist may seem to lose commercially, because patients will get to know and appreciate the honesty of the pharmacist [20].

4. The clash of economics and ethics

The effort of community pharmacists to act with integrity and adhere to the professional standards while acting in a daily business environment has been described as an act of "balancing," but such a balance seems very fragile and often threatened [25]. Reflecting on this balancing exercise between business and ethical requirements, one pharmacist expressed a good knowledge of the ethical theory, as well as of the difficulties to sticking to ethics up to the end: "You're always trying to weigh up the business versus the professional, and obviously the professional should take preference…but at the end of the day there's no point being a pauper, is there?" [24]. Vitell et al. [18] reported that 71% of their respondents (US pharmacists) considered that in their industry (pharmacy, pharmaceutical) there are a few unethical practices, whereas 13% judged such practices to be "many," a finding they interpreted that some pharmacists "are becoming more business oriented and less involved with professional activities."

The conflict between economics and ethics has been identified by pharmacy academics who, according to one source, expressed disdain towards the pharmacists from the community setting because they are employed as "graduate grocers" and jeopardize the prestige of the profession by selling cosmetics and a variety of goods that are only distantly related to healthcare if at all [3]. Some hospital pharmacists in the United Kingdom also do not have more respect for the community pharmacists, criticizing their "money grabbing, their lack of concern with professional ethics and the presentation of drugs to patients" [3, 20].

The community pharmacy (owned and operated by one or several pharmacists or large chains of pharmacies) is the setting where the business side of the profession is mostly visible and the inherent conflict between the two sides is manifest, because both pharmacy owners and employees have business-oriented responsibilities (procuring, dispensing/selling a variety of goods, managing staff) and enjoy pecuniary rewards considerably more than in other settings (especially more than in a hospital pharmacy) [9]. This may partially explain why hospital pharmacists seem to be less involved in ethical transgressions than those activating in community settings (in Australia, 1.6% of the pharmacists sanctioned for violations of professional ethics were working in hospitals and 97% in community settings) [26].

Quinney [27] interviewed 80 pharmacists (of whom 20 had been guilty of violating prescription laws or regulations) and found that 94% of the respondents provided positive answers to the following question: "Do you find that the public expects the pharmacist to be both a businessman and a professional man?"

This inner conflict between the professional and business sides of the pharmacy profession may be seen as a form of *sociological ambivalence*, a concept designating a set of norms and counter-norms, or alternating subroles, accompanying specific social positions [28], but redefined along time to describe "contradictory emotions towards the same object" (Weigert, 1991) or "the interface between individual experience and group belonging" [29]. A form of sociological ambivalence is the *ethical ambivalence*, where organizational behaviours, attitudes and norms favoured by the reward system conflict with those in line with the ethical values and judgements of the organizational stakeholders [30].

The existence of a conflict between the ethical imperative and the economic interests or pressures acting on the pharmacist may simply pass unobserved by the pharmacist. It has been reported that many such professionals have difficulties in being able to remember and relate ethical problems, a phenomenon described as "ethical inattention" or "ethical passivity [31]. The level of ethical reasoning of community pharmacists, as measured by the Defining Issues Test P% score, was reported to be lower than that of other health professions (possibly due to a process of selection and socialization), and lower levels of ethical reasoning tend to associate with lower levels of clinical performance, moral reasoning explaining a significant proportion of the variance associated with the clinical decision-making [4, 32]. However, whether this finding is generalizable to all pharmacists irrespective of place, time, educational background, etc. is doubtful, and more recent data from different countries are necessary to understand if the phenomenon does exist and what may be done in order to increase the level of moral reasoning in practicing pharmacists.

5. When business wins the battle

Contemporary codes of pharmacy ethics promote not only the classical principles of autonomy, non-maleficence, beneficence and justice, but they also tend to acknowledge that allowing business considerations to shape and dominate their demeanour may jeopardize the dignity and wellbeing of the patients [17, 33]. Although ideally ethics should outweigh business, the scarce data available till now indicate that the reverse often does happen, although the regulatory framework, culture, economic environment in which pharmacy is immersed and so on have an important role. Older research into the inherent conflict between the service and business sides of the profession and the service and commercial values reported that pharmacists for whom the business dimension of their occupation was more important than the professional one had a higher likelihood of violating the law regulating the professional activity, a finding related to the probable placing of their own interest ahead of the interests of their patients [9].

In 1854, when the pharmacy profession was still in its youth in the United States, Edward Parrish was stating that "It is mainly by the sale of quack medicines that many druggists subsist, who yet desire a reform in their business, and would be glad to co-operate in the laudable objects of the association" [34]. This illustrates the fact that well-meaning humans, including healthcare professionals and pharmacists, in particular, may have questionable practices for "subsistence."

In Germany, a complex and detailed legal regulatory framework has been (apparently successfully) adopted with the explicit purpose of ensuring that the professional aspects of pharmacy are not shadowed by commercial concerns [35]. Pioch and Schmidt [35] exploring the German pharmacy system reported that pharmacists in that country felt the strain between the "necessary evil" of the commercial side of the profession and the pharmaceutical side, with the majority of respondents (9 of 13, i.e. 69%) expressing a positive bias towards the professional side (this also means that almost one in every three pharmacists feels the bias in the contrary direction, i.e. towards business). In the words of one of the respondents about selling slimming products, "sure I would make some money if I sold it, but I also have an obligation to use my professional expertise to convince the customers that a product may be no good, even though it has been advertised" [35]. On the other hand, some of the German pharmacists (in this paper now more than 15 years old) were recognizing the increasing economic pressures in a new social and economic context, with "no rosy future" and hopes of resisting for 10–15 additional years until retirement [35].

In Australia, where the ownership of pharmacies remains the monopoly of registered pharmacists [36], a study based on interviews with community pharmacists identified economic pressures exerted over the professional activity of both employed pharmacists and pharmacist owners from community settings (unlike hospital pharmacists). Many of the employed pharmacists told of incidents where their employer (pharmacist owner) pressured them to "sell" or "move" products, whereas young pharmacist owners stated that their leading concern was not the patient's welfare but their obligation to pay back the financial institutions [17]. Financial considerations had such an impact on these latter pharmacists that the wellbeing of the patient was ignored if they perceived a risk of losing business (e.g. dispensing pseudoephedrine to a hypertensive patient) [17]. Moreover, one of the pharmacists described how the gradual increase in the business-side involvement parallels a gradual decrease in the clinical involvement, to a point where selling a product becomes more important than caring for the patient [17].

A pharmacist typology has been described that may be helpful in understanding when economics trump ethics and the other way round, but it is not clear whether this typology is innately determined, acquired or (as is more likely) determined by a mix of innate and acquired characteristics. Data from a small sample (n = 53) of pharmacists from a US middle-sized urban area including professionals from hospital and independent and chain pharmacies, based on sets of questions assessing the business- and professional-role components, identified four types of pharmacists: "business pharmacists" (in whom the business dimension is prevalent), "professional pharmacists" (the service dimension predominates), "dual pharmacists" (both dimensions have equal importance) and "indifferent pharmacists" (none of the two sides is accentuated). However, when exploring different variables potentially correlated with the different types of pharmacists, they found that both altruistic and financial values were of roughly equal importance to "business pharmacists" and "professional pharmacists," concluding that "The portrait of pharmacists torn between conflicting values is heavily overdrawn. There seems to be little support for the mutually exclusive model of service versus money and prestige values" [9]. This study used a very small sample size (with no statistical power calculations) and instruments whose validity is questionable, and thus its relevance for the understanding of the topic is limited. Quinney found a similar typology (16% professional pharmacists, 20% business pharmacists, 45% dual pharmacists and 19% indifferent pharmacists), but unlike the conclusions of Kronus, he found a disproportionate number of "business pharmacists" among the subgroup violating the regulatory framework of prescription dispensing, no "professional pharmacist" among the same subgroup, whereas those with dual or indifference orientation were represented in a lower proportion [27].

Indirect (weak) and mixed evidence regarding the influence of the spirit of capital on the professional ethics may also be obtained from other health professions. It is especially interesting how in countries with low- or middle-income and lax regulations physicians having financial links with pharmacies (such as pharmacies owned by them) tend to prescribe. A study performed in the Philippines found that physicians owning pharmacies did not tend to overprescribe, but persuaded their patients to use their own pharmacies [37]. Instead, in Taiwan, it was reported that clinics lacking on-site pharmacists had less expenditures (with

12–36%) than control sites, indicating that in such contexts physicians tended to prescribe more for financial gains [38]. In Zimbabwe, dispensing doctors (for whom there are pecuniary interests in prescribing) were reported to prescribe more medicines (including antibiotics), inject more patients and spend less time on each patient visit than non-dispensing doctors [39]. Somewhat similar results were seen in the United Kingdom, where it was found that dispensing doctors tended to prescribe more medicines and less generic drugs than the non-dispensing practices [40].

In one study published in 1991, 38% of the respondents considered that ethical standards were lower than one decade ago, whereas only 27% considered ethical standards to be superior. It is interesting to look into the views of the respondents on the most important factors causing standards to be lower or higher. Higher standards were explained by respondents through the progresses in professionalism and education (10%), the improvement of the regulatory framework (8%) and the expanded public awareness and scrutiny (3%); lower standards were explained by most respondents through economic factors: greed and the lust for profit (12%), competition and the general economic context (10%) and pricing (9%) [18]. This indicates that in the opinion of pharmacists, economic pressures are not likely to increase ethical standards, but are in all likelihood to lower them.

6. Ethics in organizational contexts

Independent exercise of the profession and autonomous ethical reasoning by pharmacists may be heavily influenced by the organizational values and objectives, oriented towards ensuring appropriate profits for shareholders [25]. Practicing pharmacy in an organizational setting (as most often does happen in practice) creates an inherent conflict between the two roles played by the pharmacist in this context of "independent professional practitioner and organizational agent" [4].

When a pharmacist is not the owner of the pharmacy in which (s)he practises, the conflict may seem to be between the ethical values of the profession and the requirements of the organization [4]; even in such cases, though, the conflict is most often of an economic nature. The adherence of the pharmacist to the organizational demands is not rooted in the inherent love for the organization but rather in the economic dependence of the pharmacist from it. On the other hand, the organization is most often interested in getting higher profits, which leads to placing drivers on pharmacists to shape their activity in line with this goal. Organizational culture and socialization may also have an impact in shaping the pharmacist behaviour as it has been argued and empirically proven that organizational systems of reward contribute substantially to the behaviours of the organization members, but rewards are often of an economic nature (although not exclusively so) [4]. The business side of pharmacy is connected —like any business—with profits, and it has been shown that reward systems of business organizations are mostly related to profits, and ethical considerations are only of a secondary importance, becoming relevant only when exposing the company to legal consequences [5].

http://dx.doi.org/10.5772/65128

In the past decades, the community pharmacy has evolved considerably in the United States, mostly from small, locally owned and managed, independent pharmacies to large, national pharmacy chains. Besides, new types of pharmacies have occurred, such as mail-order or drive-through pharmacies [33]. In a 2-year time window, it has been reported that about 3000 US independent pharmacies have given up their place to "big drug, grocery and department-store chains" [41]. In Europe, a number of countries (Belgium, Czech Republic, Poland, Romania, Switzerland and the United Kingdom) allow the operation of large retail pharmacy chains, whereas in a large number of European countries, this is not legally possible. However, in these latter countries, the multiplication of "virtual chains," associations of independent pharmacies, has accelerated over the past 10 years [42]. Even the physiognomy of third-party payers (health insurers) has substantially changed: if other times they simply paid for the work or products, they have become now more similar to corporate clients, making efforts to influence the way, form, traits and quality of pharmacy services rendered by pharmacists [3]. These metamorphoses are likely to increase the organizational socializing pressures, as well as economic constraints on pharmacists working in those pharmacies.

Organizational settings can be expected to often exert strong pressures on pharmacists to behave in certain ways (as desired by the management) and adopt particular attitudes despite different requirements of the professional code of ethics and even despite personal values and attitudes, as indicated by an increasingly large volume of empirical evidence [4]. What is more significant, this shaping of individual behaviours by organizations takes place in the absence of a "detailed set of rules, rewards, or obviously coercive structures" [43].

The approval or disapproval (normative beliefs) of employers, managers and patients was reported by Latif [44] in one study to explain only 7.6% of the variance related to clinical decision-making, but the limitations of that study suggest that organizational pressures may have a higher influence than the one measured there.

Extrapolating from other professions, Latif [51] argued that three factors may favour a selection and socialization process where those working in pharmacy may be more liable to breaching ethical standards in certain organizational settings such as the community pharmacy (this model is based on a number of empirical studies in contexts not involving pharmacists, but it is to a good extent speculative):

(a) Pharmacists with less developed ethical skills might tend to select themselves in the community pharmacy practice (positive selection).

(b) Pharmacists with more developed ethical skills might tend to leave out settings perceived as unethical (negative selection).

(c) An acculturation and assimilation of new members may lead to (various degrees of) suppression of their ethical abilities (socialization) [4].

7. Motivational outlook for choosing pharmacy

One would expect that the clash between economics and ethics is influenced by how altruistic or less so is the professional. Therefore, there is a certain interest in knowing the motivational outlook of people entering the pharmacy profession. Older studies claimed that pharmacy students entered the profession mainly because pharmacy offered the prospects of economic security or matched certain aptitudes of the candidates, whereas only a small proportion were motivated by altruistic reasons [15, 45]. More specifically, a number of first-year pharmacy students were asked about the motivations for which they selected pharmacy school and found half of them choose pharmacy because of a certain aptitude, 40% for financial security reasons and only 14% mentioned a desire to contribute the good of others [45]. This would indicate that pharmacists adopted this profession not because of altruistic reasons but for more prosaic motivations and purposes. Because these data are about five decades old, their relevance for today is difficult to establish and new investigations have to be examined. Moreover, the methodology used only provided indirect insight into the role strain, as it did not measure directly various behaviours of pharmacists.

A study on Australian pharmacy students in 2006 reported that the most important factor when choosing entry into the pharmacy school was represented by future employment prospects (somewhat confirming the older studies) and the second was "the desire to make a contribution to healthcare" (indicating a more important role to the altruistic dimension, although the question may not necessarily be interpreted in an altruistic sense) [46]. Somewhat similar results were reported in a study from 1989 in the United States [47], whereas in New Zealand, the desire in helping people was listed by students as the first reason for choosing pharmacy, while the prospect of earning a high salary was indicated by considerably less students [48]. In a recent study in United Arab Emirates, the desire to help and serve others and interest in science were found to be the most important determinants of pharmacy choice [49], whereas in Nigeria, advancement opportunities and salary were the two most important factors identified by students as reasons for entering the pharmacy schools [50]. These data (which are not exhaustive) are illustrative enough of the fact that altruistic motivations may or may not be the most important determinants of choosing pharmacy.

8. The curious relationship between age, experience and moral reasoning

Although the theory of cognitive moral development would predict that older and more experienced pharmacists have a higher level of ethical reasoning, survey-based data generated by Latif reported that first-year pharmacy students had better ethical cognition than more experienced pharmacists and moreover, the more experienced the pharmacists were, the lower their ethical cognition level was [4]. Another study carried out by the same author reported that among a group of community pharmacists, the subgroup with the lowest level of moral reasoning had a mean tenure of 22.7 years in that setting, those with a medium level had a mean tenure of 17.2 years and those with the highest level of moral reasoning had the shortest mean tenure—15.7 [51]. Because of the cross-sectional nature of the study and its limitations

[51], it is not clear whether this is a chance finding or a general truth and further investigation is necessary in this direction. Even if the findings are valid, their explanation (as proposed by Latif, who also suggested alternative explanations) may reside in the differences in educational backgrounds of the more and less experienced pharmacists, including the changes in ethics teaching along time. An older study published by Lowenthal [52] investigated in a comparative manner the answers of two groups of students (first professional year and third professional year, respectively) and one group of experienced pharmacists to certain ethical dilemmas. Unlike the findings of Latif, in this study, it was reported that both groups expressed a high priority for the welfare of the patient in the majority of situations. With respect to certain dilemmas, there were certain differences reflecting the degree of experience between the two groups, but there was no conclusion of an ethical reasoning retrogression in the more experienced pharmacists as compared with the students [52].

9. How economics may trump ethics in daily pharmacy life

The notion that (community) pharmacy is not "trade" and that commercial activities (with goods other than medicines) are only "incidental to the practice of pure pharmacy" was emphasized even in the context of the "old" pharmacy, for instance, in the United States by Wulling in 1918 [53]. However, already in 1913, cases where the business might trump ethics were identified and fought against in the literature [6]. One year later in the same line, Marshall [54] lamented in plastic words the excessive commercialization of pharmacy to the detriment of ethics: "Pharmacy has been led astray. Like the Jews of old, some of its people have set up a Golden Calf to worship and a Moses is needed to lead them back again to better, higher, if not more ethical practices."

A forceful illustration of how the spirit of capital may defeat the ethical demands in an economically advanced society is represented by selling tobacco products in the Canadian pharmacies. Although the negative impact of tobacco consumption on health is today beyond of any residual doubt and despite attempts of the professional leadership to put an end to tobacco sales in Canadian pharmacies (a movement initiated in 1985 and including several active campaigns oriented towards pharmacists), less than half of the pharmacies complied with what is an obvious ethical imperative in the first decade. As a matter of fact, in 1990 the majority of pharmacies (88%) not only sold tobacco products but even actively promoted them. As the president of the Canadian Pharmaceutical Association (CPA) acknowledged, pharmacists looked "at the bottom line," which means that they looked at the financial impact; thus, financial considerations may be in real life more important than professional demands [55]. This might be even more telling when considering that only a small minority of Canadian pharmacies were independent, whereas the majority of them belonged to large corporations, better expressions of the capital and of organizational pressures [55]. It was mainly the gradual statutory intervention of public authorities that ultimately put an end to selling tobacco products in Canadian pharmacies.

The situation was not substantially different in the United States, where community pharmacies not only benefited from selling health risk products such as tobacco and alcohol but also used advertising materials emphasizing price advantages, although such advertisements were smaller than those used for more healthful products. This was seen as an obvious case where economics tip the scales against ethics and the care for own revenues eclipses the care for the welfare of the patients [56, 57]. In one study in Massachusetts, of 100 pharmacies surveyed, 95 were selling tobacco products and half used advertisements for such products. Moreover, 42 advertised brands considered most appealing to children and 81 were willing to sell (illegally) tobacco products to minors [58]. It should also be emphasized that similarly to the Canadian situation, a minority of pharmacies (mostly independent ones) decided to place the ethics before and above economics, sacrificing the additional revenues generated by the sale of health risk product to be in agreement with their institutional mission of promoting health [56, 57]. Moreover, another US study found that pharmacists from pharmacies selling tobacco were less satisfied with their job, had more job-induced tension and had a higher proclivity to leave those pharmacies than those working in pharmacies not selling tobacco [59].

A cross-sectional study based on 377 questionnaires by pharmacists from two south-eastern US states reported that a proportion of 27% of the respondents felt sometimes conflicts between company interests and personal ethics with respect to providing information (e.g. on adverse effects of medicines) and to giving gifts and kickbacks and 1% felt often conflicts about providing information. Other conflicts between company interests and personal ethics were reported as occurring with some frequency (sometimes) regarding price collusion and pricing practices (23% of the respondents; for 2% such conflicts were frequent), honesty in executing contracts and agreements (23%), honesty in internal communications (16%), receipt of gifts and kickbacks (16%) and honesty in advertising (14%) [18]. The differences in giving and receiving gifts and kickbacks might be related to the different opportunities to give and receive gifts and kickbacks (as the authors hypothesized), but might also be related to the immoral purpose perceived more frequently in giving than in receiving such objects.

Sometimes business difficulties in pharmacy lead to fraud, as illustrated by an Australian pharmacist sanctioned by the professional body for violating professional rules in the attempt to repay business debts [26]. Any violation of the legal rules is inherently unethical, but actions and behaviours satisfying the minimal legal requirements are not necessarily ethical; the "ethical" label is to be applied to behaviours that go "above and beyond" the regulatory framework in force [5].

A study carried out in the United Kingdom, based on semi-structured interviews with seven pharmacists, reported several cases of potential conflict between economics or organizational demands and ethical demands. One of the pharmacists was troubled by the promotional activities of the "parent company," one locum pharmacist described the unease caused by the pressures exerted on them to do "what's always been done" but which was unethical, and a third one was worried that a handful of pharmacists with managerial positions would dictate in the future the ethics of the profession, based more on business than on professional considerations [24]. Thus, various behaviours may be found in pharmacy which, although not illegal, are however wide off the ethical mark.

In one of the few studies investigating the actual behaviours of pharmacists in real life, Linn and Davis [8] found that pharmacy owners were more likely to make recommendations of certain product purchases (instead of referring the patient to a physician) than were non-owners, and pharmacists working in settings losing business or with blooming business had a similar tendency. In all three cases, one might speculate that the economic interest (owners) or pressures (blooming/losing business) exert a direct influence on the professional acts.

Latif drew the attention to the reward systems in place in the community pharmacy setting that could directly collide with the professional ethics, such as rewarding prescription volumes through bonuses, a system sending the (wrong) message to the pharmacist that the volume of dispensing should prevail over patient counselling and care [51]. Unfortunately, this is not a moot consideration, but there is evidence that at least in certain places, such reward systems are implemented, as the authors of this chapter are well aware for the Romanian situation (advertisements for pharmacist jobs often mention that bonuses are conditioned on financial targets per pharmacy). In other countries, such reward systems are also likely to be in place based on our informal discussions with other pharmacists, although no systematic study seems to have ever been published in this sense.

A management oriented towards increasing profit may lead to higher workloads in pharmacy (e.g. by personnel reduction, shortening the time for patient counselling, etc.). Latif [44] reported in a questionnaire-based study that workload seem not to reduce the quality of patient care and clinical decision-making, but this study had a number of limitations, including the potential use of a workload measurement not sensitive enough, a relatively low response rate and a sample of pharmacists from a single city. Business interests may lead to "unreasonable working conditions," with a high volume of work, long working hours and no breaks, which not only are a source of frustration for the employed pharmacists [60] but also have been shown to increase the risk of errors and ethical transgressions [26]. In Australia, dissatisfied pharmacists who gave up practicing in the community setting described the working conditions and atmosphere as "unreasonable" and more similar to a factory than a professional environment, whereas in Romania, for instance, there is a widespread practice of having pharmacists working two full weekends every month, which means that a pharmacist has a full work-break only after 12 days.

In community pharmacies, the most important source of income is represented by dispensing activities. Performing activities related to pharmaceutical care (e.g. by probing deeper in the factors affecting adherence to a certain treatment scheme) takes more time and tends to erode the main revenue generating activity of dispensing [4]. Time spent in patient counselling may be seen as a waste of resources, contributing little to business objectives of the pharmacy (maximization of profit), and this has led to the opinion that high-quality patient advice may soon simply disappear from pharmacies (as illustrated by the occurrence of the mail-order pharmacies where there is no face-to-face counselling) [33]. Printed advice (such as computer printouts provided to patients together with their medications) is no substitute for the oral communication between the pharmacist professional and the patient, but quality counselling is at a high risk of becoming "a casualty in the ongoing war between pharmacy ethics and business objectives" [33]. Spending the time in patient counselling may create ethical dilemmas for each individual pharmacist, for pharmacy managers and for the top management of the pharmacy business organization, as discussed in depth by Resnik et al. [33].

The decision on providing services for drug misusers, often taken by managers, who sometimes are not pharmacists, seems to be heavily influenced by business considerations, much more than patient welfare. Such business considerations include the potential to discourage other customers/patients from entering the pharmacy, the possibility of shoplifting, the distastefulness for both staff and customers and the absence of a long-term financial gain from providing such services [25].

Other situations in which the conflict between ethics and economics may become visible include decisions by pharmacists whether to recommend or dispense unnecessary food supplements and whether to recommend lower cost generic equivalents or filling prescriptions in conditions that may be less advantageous in economic terms (e.g. for Medicaid patients) [33].

When the business side of pharmacy prevails over the professional side by an exclusive or excessive focus on profit, employed pharmacists may feel deep dissatisfaction, to the point of leaving the profession. In the words of a female pharmacist from Australia, formerly working in a community pharmacy and currently having a PhD in a non-health discipline: "I think within community pharmacy... it was very profit motivated rather than, service orientated... the job was quite isolating professionally...Very much the focus of the owner was on profit motivation rather than on, you know excellence in professional service. So very much you spend your time, obviously to please your boss" [60]. Such feelings seem not to be particular for this pharmacist, "the absence of a professional environment to work in and the challenges they face in a profit-motivated profession" being described in this study based on interviews with former pharmacists that left the profession as a "recurring topic."

10. Knowledge gaps and conclusions

It has been argued that a financial dimension exists in all professions, but in the case of pharmacy, this is more prominent, chiefly because in this case, the professional (pharmacist) sells non-pharmaceutical products in a setting that is more similar to a retail outlet than to a typical professional office [8]. In the ethical clash between business and ethics, the latter should be the winner, but day-to-day practice shows that even in countries with sophisticated and modern regulations (such as Germany), the interaction between the two remains a delicate balancing exercise. Economics may outweigh ethics in crude forms such as selling health risk products (as it was in the past for tobacco or alcohol) or in more subtle ones, such as longer work hours and higher workloads, dispensing unnecessary healthcare products (e.g. food supplements), not recommending/providing lower cost generics, etc.

It may seem surprising that empirical and theoretical research of this conflict has been so limited, when considering that its manifestation is assumed to be frequent, even daily [25]. A review of the papers published in the field of pharmacy ethics in 12 years (1990–2002) found that the volume of research carried out till now was very limited, with research on meta-ethics close to none, there was no dedicated journal of pharmacy ethics and the majority of materials published were represented by codes or statements of professional bodies, views and reflections published in manuals or debates taking place in a limited number of publications [25, 61]. Some advancement has been made in the meantime, but the overall impression when examining the available literature is not fundamentally different. It is therefore not surprising that the issue of the ambiguity of role of the pharmacist has not been explored as extensively as it deserves.

In the memorable formula of M. Brazier, there is in the academic world still too much emphasis on "ethical dilemmas of high drama and low incidence," which are of little relevance to the majority of pharmacists (as cited by Cooper et al. [62]). For community pharmacists the tension between the business and professional aspects of pharmacy might be felt considerably more often, and empirical research in this field is unjustifiably scarce. Even when available, the literature has mainly used samples of students (not fully formed or experienced pharmacists), and when pharmacists were studied, they were Anglophone, in particular from the United States, and practising outside the community pharmacy settings (chiefly hospital pharmacists) [62].

It becomes obvious then that large areas of the subject have remained not systematically investigated, whereas some of the investigations have been carried out in old times, and their relevance may be questionable today in different regulatory and economic frameworks. It would be especially useful to have data from several continents regarding the experience of community pharmacists and regarding the ways in which the conflict between economics and ethics becomes manifest and quantitative data regarding the extent of the phenomenon.

Criticizing the arguments of Denzin and Mettlin, Dingwall and Wilson insisted that the former had no empirical data on the everyday work of pharmacists and that the only evidence advanced by those authors was derived largely from "surveys, attitude studies and occupational propaganda," and emphasized the need for a research programme intended to explore the real contexts in which pharmaceutical services are provided [19]. A similar observation regarding the lack of "empirical investigations of what pharmacists actually do" had already been formulated by Linn and Davis [8]. In the United Kingdom, the Nuffield committee did not have the necessary resources for performing its own research and, noting the scarceness of the available research work, observed that what was lacking was especially "information on what pharmacists actually do—as distinct from what they say they do" [3]. After more than 30 years, things are not substantially different: most of the sparse research available in the field of pharmacy ethics is still based either on theoretical considerations or on opinions and attitudes (through qualitative or quantitative questionnaires) rather than on objective investigations of what pharmacists do in real life (although one has to acknowledge that such objective investigations are very hard to implement).

Author details

Robert Ancuceanu[1*] and Ioana-Laura Bogdan[2]

*Address all correspondence to: robert.ancuceanu@umf.ro

1 University of Medicine and Pharmacy "Carol Davila", Bucharest, Romania

2 Medone Research, Bucharest, Romania

References

[1] Harding G, Taylor K. Responding to change: the case of community pharmacy in Great Britain. Sociol Health Illness 1997; 19: 547-560. doi: 10.1111/j.1467-9566.1997.tb00419

[2] Greenwood E. Attributes of a profession. Soc Work. 1957; 2: 45–55.

[3] Dingwall R, Watson P. Small pharmacies and the National Health Service. Report for Trent Institute of Health Services Research. 2002. [Internet] Available from: https://www.academia.edu/1444036/Small_Pharmacies_and_the_National_Health_Service [Accessed: 2016-07-28].

[4] Latif D. Ethical cognition and selection-socialization. J Bus Ethics 2000;25:343–357. doi: 10.1023/A:1006097521228

[5] Hegarty WH, Sims HP. Some determinants of unethical decision behavior: an experiment. J Appl Psychol. 1978; 63: 451-457. doi: 10.1037/0021-9010.63.4.451

[6] Apple FM. The commercial advantages of properly applied professional pharmacy. J Pharm Sci. 1913; 2: 481–485. doi: 10.1002/jps.3080020413

[7] Thorner I. Pharmacy: the functional significance of an institutional pattern. Soc Forces. 1941; 20: 2321-2328.

[8] Linn LS, Davis MS. Occupational orientation and overt behaviour – the pharmacist as drug adviser to patients. Am J Public Health. 1973; 63: 502-508.

[9] Kronus CL. Occupational values, role orientations, and work settings: the case of pharmacy. Sociol Quart. 1975; 16: 171-183. doi: 10.1111/j.1533-8525.1975.tb00936.x

[10] Shaw CT. Professionalism – a sociological evaluation of commercialism and inconsistencies in pharmacy. J Am Pharm Assoc. 1971; 11: 539-544.

[11] Birenabum A. Reprofessionalization in pharmacy. Soc Sci Med. 1982; 16: 871-878. doi: 10.1016/0277-9536(82)90203-9

[12] Hibbert D, Bissell P, Ward PR. Consumerism and professional work in the community pharmacy. Sociol Health Illness. 2002; 24: 46–65. doi: 10.1111/1467-9566.00003

[13] Dessing RP, Flameling J. Ethics in pharmacy: a new definition of responsibility. Pharm World Sci 2003; 25: 3–10.

[14] Svenson G, Wood G. A model of business ethics. J Bus Ethics. 2008; 77: 303–322. doi: 10.1007/s10551-007-9351-2

[15] Denzin NK, Mettlin CJ. Incomplete professionalization: the case of pharmacy. Soc Forces. 1968; 46: 375–381

[16] Hepler CD, Strand LM. Opportunities and responsibilities in pharmaceutical care. Am J Hosp Pharm. 1990; 47: 533-543.

[17] Chaar B, Brien J, Krass I. Professional ethics in pharmacy: the Australian experience. IJPP. 2005; 13: 195–204.

[18] Vitell SJ, Rawwas MYA, Festervand TA. The business ethics of pharmacists: conflicts practices and beliefs. J Bus Ethics. 1991; 10: 295–301. doi:10.1007/BF00382968

[19] Dingwall R, Wilson E. Is pharmacy really an 'incomplete profession'? Persp Soc Probl. 1995; 7: 111-128.

[20] Holloway SWF, Jewson ND, Macon DJ. 'Reprofessionalization' or 'occupational imperialism'?: some reflections on pharmacy in Britain. Soc Sci Med. 1986; 23: 323-332. doi: 10.1016/0277-9536(86)90354-0

[21] Relman A. What market values are doing to medicine. Atlant Mthly. 1992; 269:98–102, 105–106.

[22] Shrady J. The growth of commercialism in medicine. JAMA. 1898; 30:697-701. doi: 10.1001/jama.1898.72440650005001a

[23] Fischelis RP. Commercial pharmacy or business administration? J Pharm Sci. 1922; 11: 808–809. doi: 10.1002/jps.3080111014

[24] Hibbert D, Rees JA, Smith I. Ethical awareness of community pharmacists. Int J Pharm Pract. 2000; 8:82-87. 10.1111/j.2042-7174.2000.tb00990.x

[25] Wingfield J, Bissell P, Anderson C. The scope of pharmacy ethics—an evaluation of the international research literature, 1990–2002. Soc Sci Med. 2004; 58:2383-2396. doi: 10.1016/j.socscimed.2003.09.003

[26] Penm J, Chaar BB. Professional transgressions by Australian pharmacists. J Pharm Pract Res. 2009; 39: 192-197. doi: 10.1002/j.2055-2335.2009.tb00451.x.

[27] Quinney ER. Occupational structure and criminal behavior: prescription violation by retail pharmacists. Soc Probl. 1963; 11:179-185.

[28] Merton R, Barber E. Sociological ambivalence. In Tiryalcian EA, editor. Sociological Theory, Values, and Sociocultural Change. New York: Free Press of Glencoe, 1963; p. 91-120.

[29] Hillcoat-Nallétamby S, Phillips JE. Sociological ambivalence revisited. Sociology. 2011 45: 202-217, doi:10.1177/0038038510394018

[30] Jansen E, von Glinow A. Ethical ambivalence and organizational reward systems. Acad Manage Rev. 1985; 10: 814-822.

[31] Cooper RJ, Bissell P, Wingfield J. Ethical decision-making, passivity and pharmacy. J Med Ethics. 2008;34:441–445. doi:10.1136/jme.2007.022624

[32] Latif DA, Berger BA, Harris SG, Barker KN, Felkey BG, Pearson RE. The relationship between community pharmacists' moral reasoning and components of clinical performance. J Soc Admin Pharm. 1998; 15: 210–223.

[33] Resnik DB, Ranelli PL, Resnik SP. The conflict between ethics and business in community pharmacy: what about patient counseling? J Bus Ethics. 2000; 28: 179-186. doi: 10.1023/A:1006280300427

[34] Buerki RA. The historical development of an ethic for American pharmacy. Pharm Hist. 1997; 39: 54-72.

[35] Pioch EA, Schmidt RA. German retail pharmacies: regulation, professional identity and commercial differentiation. Market Intell Plann. 2001; 19: 330-340. doi: 10.1108/EUM0000000005650

[36] Pharmaceutical Society of Australia. Pharmacy ownership [Internet]. Available from: https://www.psa.org.au/policies/pharmacy-ownership [Accessed: 2016-07-28]

[37] James CD, Peabody J, Solon O, Quimbo S, Hanson K. An unhealthy public-private tension: pharmacy ownership, prescribing, and spending in the Philippines. Health Affairs. 2009; 28: 1022-1033. doi: 10.1377/hlthaff.28.4.1022

[38] Chou YJ, Yip WC, Lee CH, Huang N, Sun YP, Chang HJ. Impact of separating drug prescribing and dispensing on provider behaviour: Taiwan's experience. Health Policy Plan. 2003; 18:316-329.

[39] Trap B, Hansen EH, Hogerzeil HV. Prescription habits of dispensing and non-dispensing doctors in Zimbabwe. Health Policy Plan. 2002; 17:288-295.

[40] Baines DL, Tolley KH, Whynes DK. The costs of prescribing in dispensing practices. J Clin Pharm Ther. 1996; 21: 343-348.

[41] Eldridge RM, Smith MF. Baker v. Arbor Drugs, Inc.: pharmacists beware of voluntarily assuming the duty to protect against harmful drug interactions. J Contemp Health L. Pol'y. 1998; 14: 41-91.

[42] PR Newswire. Over Half of Europe's Pharmacies are Grouped into Wholly Owned or Affiliated to Chains—Says New Report [Internet]. 2015. Available from: http://www.prnewswire.com/news-releases/over-half-of-europes-pharmacies-are-grouped-into-wholly-owned-or-affiliated-to-chains---says-new-report-300188788.html [Accessed: 2016-07-28]

[43] Davis-Blake A, Pfeffer J. Just a mirage: the search for dispositional effects in organizational research. Acad Manage Rev. 1989; 14: 385–400.

[44] Latif DA. Situational factors as determinants of community pharmacists' clinical decision making behavior. J Am Pharm Assoc. 1998; 38: 446–450. doi: 10.1016/S1086-5802(16)30345-X

[45] McCormack, T. The druggists' dilemma: problems of a marginal occupation. Am J Soc. 1956; 61:308-315.

[46] Davey A, Evans AM, Stupans I. Pharmacy: factors that influence the choice of career and study options. Pharm Educ. March 2006; 6: 21–26.

[47] Rascati KL. Career choice, plans, and commitment of pharmacy students. Am J Pharm Educ. 1989; 53:228–234.

[48] Capstick S, Green JA, Beresford R. Choosing a course of study and career in pharmacy —student attitudes and intensions across three years at New Zealand School of Pharmacy. Pharm Educ. 2007; 7:359–373. doi: 10.1080/15602210701673811

[49] Sharif SI, Sharif RS. Choosing pharmacy as a major: motivations and influences. Pharm Educ. 2014; 14: 116-120.

[50] Keshishian F. Factors influencing pharmacy students' choice of major and its relationship to anticipatory socialization. Am J Pharm Educ. 2010; 74:75.

[51] Latif DA. The relationship between pharmacists' tenure in the community setting and moral reasoning. J Bus Ethics. 2000; 31: 131-141.

[52] Lowenthal W. Ethical dilemmas in pharmacy practice. J Med Human Bioethics. 1988; 9: 44-49.

[53] Wulling FJ. So-called commercial pharmacy should not be taught at colleges of pharmacy. J Am Pharm Assoc. 1918; 7:872-874. doi: 10.1002/jps.3080071012

[54] Marshall EC. The commercialization of pharmacy. J Pharm Sci. 1914; 3: 1454–1455. doi: 10.1002/jps.3080031018

[55] Gibson B. Pharmacists and tobacco: dollars before duty. CMAJ. 1990; 142(6):621-622.

[56] Gupta S, Rappaport H. Advertising health risk products: ethics vs. economics. J Am Pharm Assoc. 1996; 36: 381–382.

[57] Taylor G. Pharmacists who choose not to sell tobacco. Am Pharm. 1992; 32: 49–52. doi: 10.1016/S0160-3450(15)31123-5

[58] Brown LJ, DiFranza JR. Pharmacy promotion of tobacco use among children in Massachusetts. Am Pharm. 1992; NS32(5):45-48. doi: 10.1016/S0160-3450(15)31122-3

[59] Bentley JP, Banahan BF 3rd, McCaffrey DJ 3rd, Garner DD, Smith MC. Sale of tobacco products in pharmacies: results and implications of an empirical study. J Am Pharm Assoc (Wash). 1998; 38: 703-709. doi: 10.1016/S0003-0465(15)32666-5

[60] Mak VSL, March GJ, Clark A, Gilbert AL. Why do Australian registered pharmacists leave the profession? a qualitative study. Int J Clin Pharm. 2013; 35:129–137. doi: 10.1007/s11096-012-9720-5

[61] Scharr K, Bussières JF, Prot-Labarthe S, Bourdon O. A comparative pilot study of the professional ethical thinking of Quebec pharmacy residents and French pharmacy interns. Int J Clin Pharm. 2011; 33: 974-984. doi: 10.1007/s11096-011-9570-6

[62] Cooper RJ, Bissell P, Wingfield J. A new prescription for empirical ethics research in pharmacy: a critical review of the literature. J Med Ethics. 2007; 33(2):82-86. doi: 10.1136/jme.2005.015297

Chapter 2

Rethinking Autonomy and Consent in Healthcare Ethics

Eleanor Milligan and Jennifer Jones

Additional information is available at the end of the chapter

http://dx.doi.org/10.5772/65765

Abstract

In healthcare ethics, autonomy has arguably become the 'principal principle'. As a principle that can be readily turned into a process, the giving of 'informed consent' by a patient has become the surrogate measure of whether medical interventions are ethically acceptable. While 'informed consent' processes in medical care are presumed to be robust, research confirms that most patients do not adequately understand the medical purpose, limitations or potential ethical implications of the many medical procedures to which they consent. In this chapter, we argue that the founding tenets of autonomy and informed consent which presume people to be detached autonomous individuals who act rationally from self-interest does not authentically capture the essence of human 'being'. Furthermore, such assumptions do not acknowledge the deeply relational and embedded reality of the human condition which inevitably shape decision making. We contend that within healthcare organisations, the current processes of operationalising informed consent predominantly serve legal and administrative needs, while unwittingly disempowering patients, and silencing key aspects of their experience of illness. Rather than rational self-interest, we argue that vulnerability, interdependence and trust lie at the core of ethical decision making in healthcare. Re-framing autonomy in a way that deliberately considers the unique moral frameworks, relationships, and cultures of individuals can provide a more ethically sensitive and respectful basis for decision making in healthcare. As interdependence is an integral consideration in decision making, it must be deliberately acknowledged and incorporated into healthcare practices. Embracing a narrative approach within a shared decision making framework allows the vulnerabilities, fears and aspirations of stakeholders to be heard, creating a more effective and authentic way to meet the ethical goal of respecting those who seek care.

Keywords: autonomy, consent, vulnerability, interdependence, narrative

1. Introduction

1.1. The case of Mr H

Mr H is an 83 year old Australian man who presents to the Emergency Department with a fever and in a confused state having been found wandering the streets of his neighbourhood. Mr H is known to have a pre-existing condition of Chronic Obstructive Airways Disease (COAD), and diabetes which he has been managing adequately with the help of his General Practitioner. Mr H has a 50 year old daughter, K, who lives nearby. She visits him daily, accompanies him to medical appointments, provides him with meals and attends to his cleaning and laundry needs. Mr H also has a 48 year old son, S, who lives overseas in the United Arab Emirates (UAE). Both K and S have an Enduring Power of Attorney for health matters in respect to their father. K has spoken to her brother in recent months about the need to consider alternative supported accommodation for her father as the load she carries, along with caring for family and working is becoming too great. K has experienced mild depression in recent months. S is adamant that, while his father wants to live independently and is able to make decisions for himself, their father should remain living independently. Mr H has always been clear in stating his ongoing desire to remain in his own home for as long as possible.

After admission, Mr H's confusion has been resolved on the treatment of a kidney infection. He is now medically stable although his chronic health conditions are expected to worsen significantly over time and the amount of home care needed is predicted to increase considerably. The attending team arranges a family meeting to discuss discharge planning. K and Mr H attend in person. S attends via telephone from the UAE. Mr H remains adamant that he will return home and is requesting all future care be provided in his home. He does not agree with moving into supported accommodation although the attending team is unanimous in their advice that this is the best option to meet his escalating care needs. Mr H has advised he intends to discharge himself against medical advice even though no plan has been put in place to meet his increasing healthcare needs.

Should Mr H's autonomy override all other considerations when approaching this decision? Whose decision is it anyway? What other considerations are at play? Is individual patient autonomy/choice the best lens through which to view the questions and consequences of attaining the best outcome for Mr H and his family? What is the impact of patient's choice and family conflict on healthcare staff?

In Australia, as in other Western nations, decision making in respect to healthcare matters is founded on the assumption that autonomous individual patients make choices in their best interests in alignment with their own moral frameworks. An individual is regarded as the thinking person who is able to make decisions freely, unencumbered by the needs, desires or perspectives of others. These beliefs have their origins in the work of Immanuel Kant who argued that respect for autonomy flows from '*the recognition that all persons have unconditional worth and the capacity to determine their own moral destiny*' [1]. Over many decades, our collective expectations have evolved and autonomy is now primarily regarded as individual independence.

http://dx.doi.org/10.5772/65765

One of the ways healthcare organisations have operationalized respect for this principle is through the administrative process of garnering the 'informed consent' from a patient prior to any medical intervention. The completion of an appropriately signed informed consent form, or documented discussion, is offered as assurance that any decisions or interventions undertaken, or rejected in the case of Mr H, are understood, in alignment with the patient's wishes and are ethically sound. While 'informed consent' processes in medical care are presumed to be effective, research confirms that most patients do not adequately understand the clinical purpose or personal ramifications of many interventions. They often do not understand the limitations, risks or potential ethical implications of the interventions consented to, hence, the assumption that patient 'consent' or acquiescence equates to ethically sound autonomous and informed choice should be challenged. Consequently, the basis on which informed consent processes have been founded within healthcare settings demands rethinking.

The reasons why informed consent processes have failed to substantively promote patient autonomy in healthcare are multiple. Flawed modernist conceptions of autonomy falsely elevate individualism at the expense of considering important relational considerations, the medico-legal focus of healthcare organisations can wrongly equate the minimum standard of disclosure to the higher ethical standard of respect and understanding, while environmental and educational barriers such as limited time for discussion all combine to undermine informed consent. At the core of these failures is the loss of attention to the individual patient and their vulnerability, in addition to a minimizing of the interdependent nature of healthcare.

This chapter, drawing on the case study above, challenges the dominant understandings of autonomy. It also questions how autonomy is presumed to be embedded and respected in healthcare settings. Finally, it sets the foundation for an alternative, and richer understanding of individual moral frameworks at the core of effective therapeutic relationships. With an appreciation of the vulnerability which is at the heart of the healthcare encounter, this chapter provides a way to approach respectful communication in healthcare, to understand how individuals make ethical choices, and to accept how interdependence, rather than independence, rightfully shapes the ethical tone of encounters in healthcare.

2. The rise of autonomy in medical ethics

The word autonomy is derived from the Greek words 'autos', meaning 'self' and 'nomos', meaning 'governance' or 'rule'. It is on these more individualistic understandings that informed consent procedures in the healthcare setting are premised. Manson and O'Neil [2] explain,

> *'It is easy to see why those who see autonomy as a matter of individual independence link it so closely to informed consent: informed consent procedures protect individual choice, and with it individual independence, hence individual autonomy. So if we can show that individual autonomy is a fundamental value—better still* ***the*** *fundamental value—and that it can best be protected and implemented by informed consent requirements, it may*

> *prove possible to justify informed consent procedures as required if we are to respect autonomy'* (p. 18–19).

The desire that a person should act voluntarily from a position of knowledge and understanding, to give their voluntary 'informed consent' to any medical intervention, and be supported in their capacity to exercise free will without coercion, is widely valued in Western society. Since its inception with respect to research participants outlined in the Nuremberg Code in 1947, and reinforced in subsequent versions of the Declaration of Helsinki, the specific requirement that practitioners seek patient consent has extended to clinical care. Over the past four decades, seeking consent from patients prior to medical treatment has become ingrained in health carepractice [1] that it is now *'so well entrenched that [it's] presence, indeed [it's] necessity, and justification are rarely questioned'*[i] (p. 2).

However, there is growing evidence that 'informed consent' protocols as applied in healthcare have achieved limited practical success to date[ii],[iii],[iv],[v]. Understanding the gap between the ethical intention and practical reality of informed consent requires us to revisit the philosophical origins of autonomy. The detached autonomous, individualistic 'I', who will rationally consider their accessible options and give an unencumbered and reasoned response to the question 'what ought I do', is largely a product of modernist conceptions of the self. While this concept of the 'autonomous self' permeates the bioethical discourse, as Isaacs [3] notes, *'this self is not a human self'* (p. 3); it does not reflect the lived reality of human 'being' or enhance understandings of how deeply embedded decisions are interpreted and made. People do not ignore personal sentiments in moral reasoning, nor are they detached from sentimental consequences [4]. Human beings are intentionally partial to family and friends and have mutual obligations of care, responsibility, trust and affection. Hence, while individuals may consider themselves to be rational and independent to a degree, they are unquestionably and primarily embedded and embodied beings with particular roles and responsibilities that may present non-negotiable constraints on their ability to act in a purely individualistic or rationally considered way. Social relationships, contexts and practices are not separate from autonomy or individuality. Hence, embracing the social context and accepting interdependence are integral to realising self-governance.

Other assumptions embedded within narrowly defined biomedical applications of the principle of autonomy are [5]:

- That the person making the decision, and not the community to which they belong, are the final arbiter of the integrity of the decision made.
- That real access to a variety of options actually exists, and can be readily accessed.
- That relationships with others are supportive, both psychologically and socially, and there is no consequence of abandonment or retaliation if a choice is made that others, upon whom the person may depend, oppose.

In the case of Mr H, the community at large are undoubtedly affected by his insistence that he remain at home. Such a decision is likely to require the provision of expensive individualised in-home services paid for from the communally funded healthcare system. Directing funds

http://dx.doi.org/10.5772/65765

into such care, which could be provided more economically in a shared supported care environment, may divert scarce resources from other healthcare areas where more benefit could be gained for more people. Cost must also be recognised as an ethical issue [6], hence, the decision to insist upon a certain type of care cannot be made by the patient alone. According to Baily [6] individuals do not have *'the moral or legal right to make unlimited demands…to pursue their own idiosyncratic goals'*. Health practitioners, as gatekeepers to collective resources, also have a competing moral obligation to ensure limited collective resources are spent in the most effective way to the benefit of the community more generally.

Family members arguably have duties to their ill relative; however, a relative cannot dictate, demand or impose how these presumed duties are met. In the case of Mr H, his decision to return home when he is clearly unable to attend to his own activities of daily living place a burden on K that she may be unable to fulfil. Should Mr H remain at home and K feel compelled to meet his escalating care needs, she may become further distressed and unwell. Hence, the emotional and health costs of carer burnout for K are further considerations that may create future burden on the healthcare system, and impact on K's quality of life [7]. Mr H's decision to refuse supported accommodation is likely to create friction between his son and daughter, and may undermine the psychological and social support he receives. Mr H's decision to reject the advice of his attending team may also impact negatively upon the therapeutic relationship and his ability to seek advice and care from clinicians familiar with his overall health and wellbeing. Considering all of these competing interests it is not surprising that Manson and O'Neil [2] conclude,

> *'Individual autonomy cannot be the sole principle of medical or research ethics, and consent requirements that protect individual autonomy cannot be the sole criterion of ethically acceptable action'* (p. 19).

3. The administrative and legal focus: informed consent processes and the undermining of patient autonomy

Philosophically, problems with the underlying assumptions about autonomy can derail getting to the core meanings and ethical interpretations being made by patients and practitioners alike. However, there are also administrative and legal sequelae. As the principle of respect for autonomy has become synonymous with the administrative process of garnering 'consent', informed consent processes now serve a multitude of administrative and legal (rather than ethical) purposes. According to Manson [8], these purposes include:

- Demonstrating 'proof' of ethical practice, as patient 'rights' have been incorporated into verifiable administrative processes. In some interventions, for example, there is an administrative requirement that consent forms are signed, thus ethical legitimisation is conferred to existing medical practices.
- Providing a defensive legal document lending protection to medical practitioners against potential legal recourse.

- Distancing medical care from the widely criticised paternalistic practices of the past by positioning the decisional capacity as resting solely with the patient.

Despite a wide body of social research flagging patient vulnerability and unequal power dynamics as factors which heavily compromise autonomy and weaken the capacity to freely exercise agency [9 (p. 115), 10, 11] participation alone is persistently accepted as evidence of ethically sensitive care. As any intervention taken is consensual, the belief appears to prevail that *'no injury can be done where the subject is willing'* [2] (p. 3). However, participation or willingness alone does not automatically mean that such participation is informed, nor does participation necessarily safeguard ethical acceptability, even if it does meet the minimal legal standards. Choices made from a narrow field of choices made available may not represent the most desired choice from the patients' perspective. The inevitable narrowing of choices in institutionalized healthcare may therefore conceal deeper injustices leaving them hidden and unchallenged [4, 9, 10].

The strong legal bias that informs institutionally developed consent protocols may have focused attention too heavily on facilitating the practitioner 'getting' consent, while inadequate attention has been given to supporting the full and embedded understanding of the patient 'giving' consent. Critical questions of who informs, what they inform, and for whose benefit the information disclosed can further undermine consent. While we may accept, as Beauchamp and Childress [1] do, that because *'actions are never fully informed, voluntary or autonomous, it does not follow that they are never adequately informed, voluntary or autonomous'* (p. 88), we ought to remain vigilant to institutional practices which inhibit discussion of the ethical implications of medical care. The ethical standard of understood consent cannot be presumed to be inherent within the minimal legal criteria for informed consent as the legal bias which permeates organisational practice can engulf the more subtle moral or ethical questions [11]. Indeed some commentators openly describe existing informed consent protocols as empty bureaucratic rituals [12], the main purpose of which is to provide a defensive legal document or form of insurance against malpractice suits, while paying superficial concern to the moral dimensions of care. When the motivation to procure consent becomes a matter of fulfilling a legal obligation or averting litigation, the clash of motivation between those seeking consent and those giving it potentially creates a conflict of interest that further widens the existing ethical divide between patient, practitioner and the organisation in which they meet.

Some real practical problems follow in trying to actualise informed consent in the contemporary healthcare environment. Entrenched institutional pathways such as the routinised nature of care with predetermined, expected outcomes can obscure 'choice'. Sometimes lack of genuine access to alternative options can delimit which choices are permissible. Limited time for consultations can prevent the formation of effective communication partnerships between the patient and their family, while poor access to adequate education and knowledge building converge to create a situation in which a patient's individual agency can become significantly compromised.

Against this backdrop it is perhaps not surprising that low levels of informed consent persist across many fields of medical intervention. Disturbingly low figures in meeting the ethical requirements of informed consent have been reported at 0.5% success for *'complex'* decisions,

http://dx.doi.org/10.5772/65765

defined as having extensive effect on the patient with uncertain and multiple outcomes, and reaching a mere 26% success for simple decisions where the effect is minimal and the outcomes are clear and singular [13]. Such poor success in achieving informed consent within medical care generally exposes the failure of current approaches. Although the accepted protocols may meet the minimum legal or administrative requirements, they commonly struggle to fulfil their ethical purpose of preserving patient autonomy and empowering the patient with the right to intentionally embrace or reject the interventions interventions [11, 14–16].

Collectively, these administrative and legal considerations can act to undermine patient understanding and choice. With respect to Mr H, his family and attending team—if the team simply takes Mr H at his word, the minimal legal and administrative requirements can be fulfilled by explaining the risks of discharge against medical advice and documenting this in Mr H's chart. As an autonomous adult with capacity to decide, Mr H has the right to make decisions that may not be in his best medical interests as defined by others. He can also make decisions that create burden for others, for example, a burden on his daughter K and on the shared healthcare resources of the district. Should Mr H deteriorate further at home due to inadequate care, the cost of his next admission is likely to escalate as the acuity of his medical needs increase [17, 18]. The team may feel they have respected the autonomy of Mr H, and they can confidently watch Mr H go home to let the inevitable decline unfold. As they have no direct duty of care to K or S, they need not consider the impact beyond Mr H. They have fulfilled their legal and administrative requirements. Yet, this outcome may not be seen as satisfactory from an ethical perspective as the harms that will inevitably flow to many stakeholders from such a decision have arguably not been thoroughly considered. This outcome also denies Mr H the chance to begin a dialogue about endof-life planning, a decision which may ultimately place a higher burden on his family and the healthcare service.

4. A different view: vulnerability, interdependence and narrative as integral to ethical healthcare

The practical question of whether Mr H's outcome could be different if we took a different view of autonomy begs consideration. If we embraced this different view and allowed it to permeate our administrative and educational processes in healthcare, how might patient, carer and health practitioner relationships and decisions be affected?

We propose that there is a deeper, more realistic way of thinking about human beings, in particular vulnerable human beings, in the context of healthcare. Philosophers such as Taylor [19], and many feminist scholars [20–23] have written extensively outlining such an alternate reframing of the self as deeply relational and socially embedded. Recognising that the '*individual is whole only in a world of others*' [24] the dominant conception of autonomy as overly individualistic, atomistic and detached is rejected.

Hoffmaster [25] further considers the consequences of assuming human beings are first and foremost detached and rational creatures.

> *'Human beings are rational, but human beings also have bodies and because they have bodies, they are vulnerable. In fact vulnerability is an even more basic feature of our human constitution than rationality, because while all human beings are vulnerable, not all are rational or even possess the potential to become rational…it is our very vulnerability that creates the need for morality…vulnerability marks the limits of individualism…'* (p. 43).

Acknowledging patient vulnerability creates the ethical imperative to care. It is arguably vulnerability rather than patient autonomy that shapes the moral core of clinical practice [26–29]. In the clinical setting, vulnerability and interdependence defines the patient experience creating a moral imperative for trustworthy others to care. Together vulnerability, interdependence and trust form the moral essence of healthcare.

As human beings are primarily relational beings, our experience is principally defined by social connectedness embedded in relationships and power dynamics. Each 'autonomous' act has repercussions beyond the immediate decision maker. Additionally, broader social contexts and practices may support or inhibit individual choice. Relationship and interdependence are undeniably key elements of autonomy, hence acknowledging, incorporating and respecting these must form a central foundation of healthcare delivery. Current informed consent protocols distance these insights preferring instead a model which seeks to de-contextualise and detach ethical deliberations, then re-insert a solution as if it were independent of the uncertainty and complexity which created it.

Taylor [19] maintains that the human self is not primarily a rational thinker, but rather an actor or doer, projected into a complex world in which they seek to make and find meaning. Capable of interpreting their world and their place in it, human beings are also capable of creating and re-creating, interpreting and re-interpreting themselves and others. Therefore, a unique aspect of being human is the capacity to change, grow and become when given the support and opportunity to do so. An individual's moral knowledge is not static but constantly evolving and subjectively interpreted. If we see the fundamental purpose of informed consent as a practical exercise of ensuring the needs, interests and aspirations of all participants are respected, then consent must be founded on shared appreciation of people as related, dialogical, and interpretative beings who are embedded in multiple contexts of history, culture, language, relationships, biology, time and spiritual horizons entrenched in complex social dynamics. Each of these layers informs the meaning of illness and shapes the choices made. Appreciation for the complexity of meaning making must therefore be actively sought in the healthcare environment.

With this view of the self at the centre of human interactions, a narrative approach to decision making in healthcare enables us to enter into, and appreciate, each other's understandings with conversation and dialogue providing an appropriate entry portal into the world of another's experience and knowledge. Indeed, Taylor [19] contends, *'the nature of our language and the fundamental dependence of our thought on language makes interlocution in one or other of these forms inescapable for us'* (p. 38). What is vitally important is that the conversation begins and ends within the deeply contextualized relationships between staff, patients and the organization.

4.1. The role of narrative in healthcare

In considering how human beings construct meaning from the experiences that weave the fabric of their everyday lives, Bruner makes the point, '*we organise our experience and our memory of human happenings mainly in the form of narrative—stories, excuses, myths, reasons for doing and not doing, and so on*' [30] (p. 4).

Narrative invites us to enter into the being and becoming of others, to appreciate their embodied and embedded experiences. It opens the possibility of sharing, understanding and appreciation of the complex and unique lived reality and plural moral frameworks that are unique to each person. It further enables the expression of experiences of loss, suffering, or oppression to be voiced, especially by those who may feel disempowered or silenced. Finally, narrative opens up new meanings, new possibilities and new sensitivities for collective and individual moral insight and growth.

Due to the dialogical nature of being human, narrative is widely recognised as significant in enhancing ethical understandings [31–35]. By voicing our experiences in dialogue with another, our shared common understandings are deepened which in turn opens each person to the rich complexities of each other's lives. Through this deepening of shared understandings and ethical orientations/motivations, our individual and shared ethical judgments can be better informed. Thus narrative and dialogue are pivotal in shaping not just the ethical agenda, but also the process of *fully* informed decision making [36, 37]. Kearney [38] notes,

> *Far from being ethically neutral, each story seeks to persuade us one way or another about the evaluative character of its actors and their actions… stories alter our lives as we return from text to action. Each story is loaded* (p. 55–56).

According to Canadian medical sociologist Arthur Frank [39], narrative has a further role to play in illness. Narratives provide us with the means for not only making sense of the disruption that illness has wrought on our lives but also is the means by which we are able map and re-map the future direction of our life. Stories '*repair the damage that illness has done to the ill person's sense of where she [sic] is in life, and where she [sic] may be going*' (p. 53). As the means by which both the path and the destination can be mapped, *fully* informed consent must therefore take a narrative form if the differing perspectives of each of the characters who make up any one life story are to be accounted for. Frank further contends that illness narratives are shaped around one of three narrative forms: the restitution narrative in which the ill person seeks to return to a pre-ill state; the chaos narrative in which the ill person cannot imagine life improving; and the quest narrative in which the suffering is met head on—the illness (and accompanying suffering) is accepted and is used for improvement. Each person's narrative is deeply embedded in another's. Therefore, a change in the shape of one's life narrative inevitably impacts on both the life and narrative of each relational 'other'. It is this truth of human interaction that compels us to rethink autonomy and place relationship at its core.

If Mr H's story was pieced together in the coherent whole from his perspective, what might his children and the attending team come to understand? Might his identity be strongly interlinked with being in his house where he keeps the memories of his wife and their life together alive? Might the routines he has established over many years be a strong comfort to him? If K could voice her story might she express signs of carer fatigue and need opportunities for respite, or might she reveal her sense of life purpose is in tension with her father, brother, husband and children? Could S's story reveal something other than being autocratic and distant from the day to day realities of his father? Might he have some legitimate concerns for his father being moved not previously taken into account? Might the deeper understandings of the attending team resulting from being engaged in many similar stories of people in Mr H's position with his condition offer insights for Mr H and his children? Voicing and listening attentively together to these diverse narratives about Mr H's current and future life may build up a new narrative. Such a new narrative might better integrate and unify the values each viewpoint is motivated by and may enable the creation of a story all can find morally beautiful:

> *Giving voice to the moral beauty we recognise in others is a loving and caring act that is not without its influence on our sense of self. The self always exists in relation to others and is authored in conversations with many others. The self flourishes in a community of dialoguers giving voice to moral beauty* [40].

Finally, as Connelly [41] notes, creating strong therapeutic relationships must begin with 'mindfulness', an openness to help the patient tell their story and a commitment to value this unique story when it is told.

> *'Physicians must be able to help the patient tell the story that is most important, meaningful and descriptive of the situation. If the patients' narrative is not heard fully, the possibility of diagnostic and therapeutic error increases, the likelihood of personal connections resulting from a shared experience diminishes, empathic opportunities are missed and patients may not feel understood or cared for'* (p. 84).

5. Education and communication

The current focus on 'getting' informed consent to meet legal and administrative requirements has fuelled approaches to patient education in healthcare that can be superficial and often woefully inadequate. Commonly, research indicates that educational materials aimed at assisting patients in giving informed consent frequently omit topics of relevance to patients, overestimates the usefulness and benefits of intervention, and often contain inaccurate, misleading, or incomplete information [42–48]. Coulter's [44] review of informed consent practices found widespread failure to provide accurate and balanced information of relevance to patients entrenched across many fields of medical care. This research concluded:

> *'Current information materials for patients omit relevant data, fail to give a balanced view of the effectiveness of different treatments and ignore uncertainties… the most common fault [of educational material] was to give an overoptimistic view, emphasising the benefits and glossing over the risks and side effects'* [44] (p. 318).

As a basis for informed decision making, promoting patient understanding is critical to respecting their choices. When sound understanding is achieved within a strong therapeutic partnership of mutual contribution and respect, research shows positive outcomes for patients

http://dx.doi.org/10.5772/65765

and practitioners including active acceptance of outcomes (positive and negative), increased satisfaction and more stable decisions, higher compliance with treatment, lower instances of litigation and fewer ongoing referrals [49–52]. Effective communication and education are therefore critical in building enduring and effective therapeutic alliances.

Within a framework of relationally aware, socially contextualized and interdependent understandings of autonomy and informed consent, practical problems of how to inform (in addition to 'what' to inform) arise. The further question of how to translate information to understanding that has meaning for the person making the decision must also be considered. This is essentially a process of learning and education. The importance of incorporating successful teaching, learning and communication strategies into strengthening informed consent procedures is a practical route to explore. Such an educational stance in the clinic requires explicit recognition of the multiple underlying social, personal, institutional and philosophical constraints that impinge upon the practical realisation of autonomy, consent and choice.

5.1. Communication: commitments and barriers

> *'Strangers taking care of strangers can come to be as pathological a factor in certain situations as sepsis'* [12] (p. 95).

Dialogue and communication lie at the ethical core of human interactions in healthcare. The ability of patients and physicians to communicate effectively is an essential element of medical care, as it is through these dialogical encounters that mutual appreciation and understanding can emerge. Consequently, barriers to effective communication contribute significantly to the ethical failures observed both in the existing literature as they render patients and practitioners vulnerable. Time pressures, under-resourcing, inadequate support and/or training of staff and, perhaps, organisational, institutional and gendered hierarchies of power in which the patient-clinician encounter is entrenched can all significantly undermine effective communication.

While it is generally acknowledged that effective communication lies at the ethical core of clinical encounters, research into communication patterns between healthcare providers and patients predominantly paints a bleak picture [53]. Additionally, the well catalogued tendency to position ethical dialogue and deliberation within carefully demarcated 'dilemmas' or 'headliner quandaries' [54] has arguably narrowed our view of what constitutes an 'ethical' conversation. In fact, every encounter is an ethical one, premised on vulnerability, interdependence and trust, on the *'micro-ethical'* [55] or underlying *'moral sensibilities'* [56] of everyday practice, injustices, and relationships. Komesaroff [55] observes, from his perspective as a physician, that ethics not a technical expert discussion, but rather, ethics is *'what happens between every patient and every doctor everyday'* (p. 68). Komesaroff [55] further notes that *'the vast majority of clinical decisions are taken in an ethical environment in the absence of any obvious dilemma'* (p. 67). The educational and professional conditioning of practitioners to regard the ethical within a limited 'dilemma' or 'quandary' framework ultimately impoverishes their human understandings and thus fails to equip them for the ongoing process of guiding morally sensitive interactions. When ethical debate and education focuses on such remote or extreme examples, an insidious consequence is that the morality of the commonplace—which is the heart of the ethical—becomes marginalized. Kass [56] explains:

> *'The use of surrogate wombs or the definition of death or guidelines for terminating life sustaining treatment captures most of the attention- not surprisingly—but the morality of ordinary practice is largely ignored. Yet every encounter is an ethical encounter, an occasion for the practice and cultivation of virtue and respect, and between doctors and patients for the exercise of responsibility and trust on both sides. How do physicians speak to patients? Do they have reasonable expectations of their physicians? How do we, individually and culturally, stand with respect to rearing children, sharing intimacies, revering life, facing death? In the absence of attention to these more fundamental and pervasive moral postures and practices, is it reasonable to expect that an ethics for the extreme cases will be sensibly worked out even in theory, let alone be successful when 'applied' to practice?'* (p. 2).

Thus, there is a strong need for practitioner ethics education at every level: graduate; undergraduate and ongoing specialist training to be 're-humanised'. Such re-humanisation must prepare practitioners with the necessary hermeneutical insights and sensitivities to become responsive ethical partners and advocates attuned to the unique everyday moral sensibilities and experiences of their patients. In essence, healthcare ethics must be taught as a transformative, relational, engaged and fundamentally human endeavour. It must be taught as a discipline that raises sensitivity to the vulnerable, instils a broader view of the sources of suffering and harm experienced by patients, and a self-awareness of one's own values and beliefs and how they impact interactions with others.

The 'medical humanities' within healthcare ethics education provides some practical leadership in enhancing practitioners' humanistic and ethical understandings of themselves and their patients. Similarly grounded in a hermeneutical view of the ethical, the medical humanities invite exploration of individual illness experiences through creative works such as literature, poetry, music, biography and art. Thus, we are invited to consider the unique experience of each person, as opposed to viewing the ethical response in terms of '*general principles and typical patients*' [57] (p. 127).

The goal of patient self-determination permeates healthcare rhetoric. For the most part, patients are known to welcome their shared decision making role [49, 58–68]. Embracing 'patientcentred' care, and shared decision making, better supports patient autonomy and is also regarded as a more appropriate and respectful basis from which to form strong therapeutic partnerships. While there is no definitive definition of shared decision making, it is described as an intentional mode of patient/practitioner interaction defined by a partnership of mutual respect, of equally valuing patient preferences, of discussing all options, benefits and risks, of facilitating appropriate education, and of negotiation, deliberation and seeking of mutual agreement on healthcare decisions [69].

The adoption of shared decision making within a patient-centred model of care has many positive health outcomes such as reducing excessive diagnostic testing and ongoing referrals, lowering the incidence of malpractice suits, reducing anxiety, increasing patient adherence to

http://dx.doi.org/10.5772/65765

treatment regimes, improving health literacy of patients and generally improving the health status of individuals [49–51, 70]. Such positive outcomes may be directly attributable to patient perceptions of being included as a full participant [71] in the clinical encounter rather than as a mere passive recipient of treatment or imposed decision. It also has an ethical impact by acknowledging the patient's contribution and expert knowledge.

6. Conclusion

Respect for autonomy as constructed in dominant accounts within bioethics and implemented through informed consent processes in healthcare has failed to deliver on its goal of valuing individuals and promoting individual choice. The legal and administrative processes of garnering 'informed consent' have arguably marginalized patients and silenced key elements of their experiences and goals. As interconnection, relationship, vulnerability and trust are the core elements of human interaction and understanding in healthcare these elements, rather than detached individual choice, must form the basis of practitioner-patient interactions in healthcare. This can be achieved by embracing a narrative approach to therapeutic relationships which enable each stakeholder to be heard, to build shared understanding and to navigate the path to ethical care which respects the embedded and unique moral framework and illness experience of each person. This approach opens access to a richer and more realistic account of autonomy.

Author details

Eleanor Milligan[1*] and Jennifer Jones[2]

*Address all correspondence to: e.milligan@griffith.edu.au

1 Griffith University School of Medicine, Queensland, Australia

2 Princess Alexandra Hospital, Brisbane, Australia

References

[1] Beauchamp T, Childress J. Principles of biomedical ethics. 5th ed. New York: Oxford University Press; 2001.

[2] Manson NC, O'Neill O. Rethinking informed consent in bioethics. Cambridge: Cambridge University Press; 2007.

[3] Isaacs P. *Ontology, narrative and ethical engagement*.In Confessions: Confounding Narrative and Ethics, 2010. Edited by Eleanor Milligan and Emma Woodley. Newcastle upon Tyne: Cambridge Scholars Publishing; 2010.

[4] Sherwin S. No longer patient. Feminist ethics and health care. Philadelphia: Temple University Press; 1992.

[5] Oshana MAL. The autonomy bogeyman. Journal of Value Inquiry 2001;35:209.

[6] Bailey MA. Utility, autonomy, and cost in end-of-life care. The Journal of Law, Medicine and Ethics 2011;39:172–82.

[7] Kasuya R, Polgar-Bailey P, Takeuchi R. Caregiver burden and burnout: a guide for primary care physicians. Postgraduate Medicine 2000;108:119.

[8] Manson NC. Consent and informed consent. In: Ashcroft RE, Draper H, McMillan JR (Eds.) Principles of health care ethics. 2nd ed. New York: John Wiley & Sons Ltd; 2007.

[9] Sherwin S. Diagnosis difference: the moral authority of medicine. Hypatia 2001;16:172.

[10] Milligan, E. The ethics of consent and choice in prenatal screening. Newcastle upon Tyne: Cambridge Scholars Publishing; 2011.

[11] Wear S. Informed consent: patient autonomy and clinician beneficence within health care. 2nd ed. Washington D.C.: Georgetown University Press; 1998.

[12] Wear S. Informed consent: patient autonomy and clinician beneficence within health care. 2nd ed. Washington D.C: Georgetown University Press; 1998.

[13] Braddock CH, Edwards KA, Hasenberg NM, Laidley TL, Levinson W. Informed decision making in outpatient practice. Time to get back to basics. The Journal of the American Medical Association 1999;282:2313–20.

[14] Braddock CH. Factors that predict better informed consent. Journal of Clinical Ethics 2002;13:344.

[15] Braddock CH, Edwards KA, Hasenberg NM, Laidley TL, Levinson W. Informed decision making in outpatient practice. Time to get back to basics. The Journal of the American Medical Association 1999;282:2313–20.

[16] Bottrell M, Alpert H, Fischbach R, Emanuel L. Hospital informed consent for procedure forms. Archives of Surgery 2000;135:26.

[17] Chan WC, Jackson G, Winnard D, Anderson P. Healthcare services funded by Counties Manukau District Health Board for people in the last year of life. The New Zealand Medical Journal (Online) 2011;124:40–51.

[18] Garden G, Green S, Pieniak S, Gladman J. The Bromhead Care Home Service: the impact of a service for care home residents with dementia on hospital admission and dying in preferred place of care. Clinical Medicine 2016;16:114–8.

http://dx.doi.org/10.5772/65765

[19] Taylor C. Sources of the self. Cambridge: Harvard University Press; 1989.

[20] Friedman M. Autonomy and social relationships. Rethinking the feminist critique. In: Meyers DT (Ed.) Feminists rethink the self. Oxford: Westview Press; 1997.

[21] Donchin A. Understanding autonomy relationally: toward a reconfiguration of bioethical principles. Journal of Medicine and Philosophy 2001;26:365–86.

[22] Dodds S. Choice and control in feminist bioethics. In: CMNS (Ed.) Relational autonomy. New York: Oxford University Press; 2000, pp. 213–35.

[23] Sherwin S. The politics of women's health: exploring agency and autonomy. The Feminist Health Care Ethics Research Network. Philadelphia: Temple University Press; 1998.

[24] Cassell, E. Recognizing Suffering. The Hastings Center Report. 1991: 21.3, pp 24–31.

[25] Hoffmaster B. What does vulnerability mean? The Hastings Center Report. 2006;36:38.

[26] Lanoix M. The ethics of imperfect cures: models of service delivery and patient vulnerability. Journal of Medical Ethics 2013;39:690.

[27] Martin AK, Tavaglione N, Hurst S. Resolving the conflict: clarifying 'vulnerability' in health care ethics. Kennedy Institute of Ethics Journal 2014;24:51–72.

[28] Hurst SA. Vulnerability in research and health care; describing the elephant in the room? Bioethics 2008;22:191–202.

[29] Tavaglione N, Martin AK, Mezger N, et al. Fleshing out vulnerability. Bioethics 2015;29:98–107.

[30] Bruner J. The narrative construction of reality. Critical Inquiry 1991;18:1–21.

[31] Walker MU. Keeping moral space open: new images of ethics consulting. The Hastings Centre Report. 1993;23:33–40.

[32] Walker MU. Moral understandings: a feminist study in ethics. London: Routledge; 1998.

[33] Tomlinson T. Perplexed about narrative ethics. In: Lindeman-Nelson J (Ed.) Stories and their limits. New York: Routledge; 1997.

[34] Brody H. "My story is broken can you help me fix it?" Medical ethics and the joint construction of narrative. In: Fulford KWM, Dickenson DL, Murray TH (Eds.) Healthcare ethics and human values: an introductory text with readings and case studies. Malden, Massachusetts: Blackwell Publishers Ltd.; 2002.

[35] McMillan J, Gillett G. Consent as empowerment: the role of postmodern and narrative ethics. In: Fulford KWM, Dickenson DL, Murray TH (Eds.) Healthcare ethics and human values: an introductory text with readings and case studies. Malden, Massachusetts: Blackwell Publishers Ltd.; 2002.

[36] Taylor C. Sources of the self. Cambridge: Harvard University Press; 1989.

[37] Widdershoven GAM, Smit MJ. Ethics and narratives. In: Josselson R (Ed.) Ethics and process in the narrative study of lives. Thousand Oaks: Sage Publications; 1996.

[38] Kearney R. On stories. London: Routledge; 2002.

[39] Frank A. The wounded storyteller: body, illness and ethics. Chicago: University of Chicago Press; 1995.

[40] Hancock DW. An exploration of the virtues project: ontological, educational and cross-cultural inquiries into a moral education program within a Mongolian school. Australia: Queensland University of Technology; 2015.

[41] Connelly J. Narrative possibilities. Using mindfulness in clinical practice. Biology and Medicine 2005;48:84–94.

[42] Coulter A. Evidence based patient information. British Medical Journal 1998;317:225–6.

[43] Coulter A, Entwistle V, Gilbert D. Sharing decisions with patients: is the information good enough? British Medical Journal 1999;318:318-22.

[44] Coulter, A What do patients want from high quality general practice and how we involve them in improvement? British Journal of General Practice 2002: 52 (Suppl); S22–S26.

[45] Lawson KL. Expectations of the parenting experience and willingness to consider selective termination for Down syndrome. Journal of Reproductive and Infant Psychology 2006;24:43.

[46] Williams C, Alderson P, Farsides B. What constitutes 'balanced information in the practitioners' portrayal of Down's syndrome? Midwifery 2002;18:230–7.

[47] Bramwell R, Carter D. An exploration of midwives and obstetricians knowledge of genetic screening in pregnancy and their perception of appropriate counselling. Midwifery 2001;17:133–41.

[48] Burkell J, Campbell GD. "What does this mean?" How web-based consumer health information fails to support information seeking in the pursuit of informed consent for screening test decisions. Journal of the Medical Library Association 2005;93:363.

[49] Stewart M, Belle-Brown J, Donner A, et al. The impact of patient—centred care on outcomes. The Journal of Family Practice 2000;49:796–804.

[50] Stewart M. Effective physician-patient communication and health outcomes: a review. Canadian Medical Association Journal 1995;152:1423–33.

[51] Levinson W, Roter DL, Mullooly JP, Frankel RM. Physician-patient communication: the relationship with malpractice claims among primary care physicians and surgeons. The Journal of the American Medical Association 1997;277:553–9.

http://dx.doi.org/10.5772/65765

[52] Bekker HL, Hewison J, Thornton JG. Understanding why decision aids work: linking process with outcome. Patient Education and Counselling 2003;50:323–9.

[53] Braddock CH. Factors that predict better informed consent. Journal of Clinical Ethics 2002;13:344.

[54] Liaschenko J, Oguz NY, Brunnquell D. Critique of the "tragic case" method in ethics education. Journal of Medical Ethics 2006;32:672–7.

[55] Komesaroff, P. From bioethics to microethics: ethical debate and clinical medicine. In Komesaroff, P.A. (Ed.) Troubled bodies: critical perspectives on postmodernism, medical ethics, and the body, Melbourne: Melbourne University Press.

[56] Kass LR. Life, liberty and the defense of dignity. The challenge for bioethics. San Francisco: Encounter Books; 2002.

[57] Begley, A. Creative approaches to ethics: prose, poetry and dialogue. In Tschudin, V. (Ed.) Approaches to ethics: nursing beyond boundaries. London: Elsevier Science, 2003.

[58] Godolphin W, Towle A, McKendry R. Challenges in family practice related to informed and shared decision-making: a survey of preceptors of medical students. Journal of Applied Mathematics and Computing 2001;165:434–8.

[59] Bub B. The Patient's lament: hidden key to effective communication: how to recognise and transform. British Medical Journal 2004;30:63–9.

[60] Fisher B, Dixon A, Honeyman A. Informed patients, reformed clinicians. Journal of the Royal Society of Medicine [NLM-MEDLINE] 2005;98:530.

[61] Stewart M, Belle-Brown J, Weston W, McWhinney I, McWilliam C, Freeman T. Patient-centred medicine: transforming the clinical method. 2nd ed. Oxon: Radcliffe Medical Press; 2003.

[62] Towle A, Godolphin W. Framework for teaching and learning informed shared decision making. British Medical Journal 1999;319:766–71.

[63] Weston W, Belle-Brown J. The importance of patients beliefs. In: Stewart M, Roter D (Eds.) Communicating with medical patients. Newbury Park: Sage Publications; 1989.

[64] Weston W. Informed and shared decision making: the crux of patient centred care. Canadian Medical Association Journal 2001;165:438–9.

[65] Elwyn G, Edwards A, Kinnersley P. Shared decision making in primary care: the neglected second half of the consultation. British Journal of General Practice 1999;49:477–82.

[66] Say R, Thomson R. The importance of patient preferences in treatment decisions-challenges for doctors. British Medical Journal 2003;327:542.

[67] Little P, Everitt H, Williamson I, Warner G, al e. Preferences of patients for patient centered approach to consultation in primary care: observational study. British Medical Journal 2001;322:468.

[68] Coulter A, Magee H. The European patient of the future. Maidenhead: Open University Press; 2003.

[69] Makoul G, Clayman M. An integrative model of shared decision making in medical encounters. Patient Education and Counseling 2006;60:301–12.

[70] Braddock CH. The emerging importance and relevance of shared decision making to clinical practice. Medical Decision Making 2010;5S–7S.

[71] Stewart, M. Belle-Brown, J. Donner. A. The impact of patient-centred care on outcomes. The Journal of Family Practice, 2004; 49.9: 796–804.

Chapter 3

Ethical Publications in Medical Research

Kusal K. Das and Mallanagoud S. Biradar

Additional information is available at the end of the chapter

http://dx.doi.org/10.5772/64947

Abstract

Ethics in medical sciences research may not always translate into ethical publications. Unfortunately due to lack of regulatory bodies, publication misconduct is now a global menace for the scientific community. Publication misconducts are not only restricted to research fraud or data manipulations alone but also seriously include plagiarism, duplicate publications especially on figures and tables, authorship disputes and conflict of interests. As global scientific research is expanding particularly in the field of health sciences hence possibilities of more rise of unethical practices from research to publications are very high, authors suggest a strong peer-reviewing system, use latest technological support, strong publication ethics policies, active monitoring, protection of whistle blowers and more liaisons between journals and research institutions or universities possibly to prevent publication misconduct effectively. This chapter discusses how medical publications might have abused various ethical norms not only while conducting research but also during the publication process. The review also discusses the possible preventive measures against unethical practices of research publications.

Keywords: scientific misconduct, medical journals, ethical publications

1. Introduction

Ethics in medical sciences research may not always translate into ethical publications. As peer pressure rises the ethics of conducting medical research and subsequent writing scientific papers and publications gradually erodes in the last couple of years. This phenomenon so much deeply penetrates into the medical researchers that various professional bodies, universities and governments are forced to press panic button against unethical medical research and publications [1]. Ethical violations in conducting medical research always promote unethical scientific publications. The most important outcome of any research is its findings and observations and definitely improper research or scientific misconduct will lead to unethical publications. The research misconduct that promotes unethical publication impacts badly on other researchers who follow the steps shown in unethical scientific publications and resulting wrong practices or applications on patients [2]. Scientific and research misconduct is defined very clearly by the Royal College of Physicians at Edinburgh —'as any behaviour by a researcher, whether intentional or not, that fails to scrupulously respect high scientific and ethical standards. Various types of research misconduct include fabrication or falsification of data, plagiarism, problematic data presentation or analysis, failure to obtain ethical approval by the Research Ethics Committee or to obtain the subject's informed consent, inappropriate claims of authorship, duplicate publication and undisclosed conflict of interest (COI)' [1]. The statement specifically mentioned that research misconduct does not end at the research works level but also extends to the publication level. One must note that research misconduct either is done intentionally or unintentionally—hardly it matters on its impact to the society that includes fellow researchers, authors, reviewers, editors, institutes, universities, nations and above all future students of medicine, professionals and patients as a whole. In the era of 'publish or perish' medical fraternity should not focus only on his/her career advancement but also consider the professional ethics including research and publication ethics seriously [3]. How serious a research misconduct may be the story of South Korean stem cell scientist Woo Suk Hwang is enough to speak to that! Dr. Hwang's revolutionary work on stem cells published in Science (2004 and 2005) and later found that both the papers are fakes [4].

According to Fanelli, research misconduct should be redefined as 'any omission or misrepresentation of the information necessary and sufficient to evaluate the validity and significance of research, at the level appropriate to the context in which the research is communicated' [5]. Faneli also stated that 'scientific knowledge is reliable not because scientists are more clever, objective or honest than other people, but because their claims are exposed to criticism and replication' [5].

The consequence of research misconduct not only tarnishes the image of the spirit of science but also collaterally damages many things like:

1. Society and humanity: Wrong procedures, false and fabricated data bring out products, which may be considered unsafe for humanity. Here comes the publication ethics regulation, which perhaps control or prevent these danger.

2. Fellow researchers: Published data and knowledge derived from research misconduct in medical sciences will mislead fellow medical researchers and that will lead to huge loss of money, funds, times and reputations.

3. Medical practitioners and students: Medical practitioner also suffers a lot due to unethical research publications as many wrong diagnostic and therapeutic published guidelines lead to professional disaster for them. Medical students might be taught subjects and understanding based on false and fabricated data which will jeopardize the career of future doctors.

4. Public trust and Government policies: Research misconduct and subsequent unethical publications may destroy public trust on science. Such false information and data may misguide government and lead to implement some erroneous health policies and laws. The ultimate sufferers are common man and society.

Hence, we can say that all the stakeholders from researchers, institutions/universities, government agencies, medical journals or book publishers are going to be devastated by research misconduct, which may also be considered as the most serious scientific assault on human health sciences.

While conducting medical research, researchers are usually careful and take all the precautions against any sort of ethical violation either in human or in animal research as per the guidelines of various apex professional bodies. Institutional Regulatory Body/Institutional Ethical Committee of all the countries function near similar pattern which strictly follow Declaration of Helsinki and other international guidelines. In general, all the institutional ethics committee critically obey all the ethical principles as per the Declaration of Helsinki by World Medical Association (WMA), National Institute of Health (NIH), the Food and Drug Administration (FDA), Environmental Protection Agency (EPA), Singapore statement of research integrity, ICMR guidelines, etc. In a nutshell, American Psychological Association comes out with five principles for research ethics: (i) discuss intellectual property frankly, (ii) be conscious of multiple roles, (iii) follow informed consent rules, (iv) respect confidentiality and privacy and (v) tap into ethics resources [6]. NIH also summarizes the principles of 'Codes and Policies for Research ethics' as the following: (i) honesty, (ii) objectivity, (iii) integrity, (iv) carefulness, (v) openness, (vi) respect for intellectual property, (vii) confidentiality, (viii) responsible publication, (ix) responsible mentoring, (x) respect for colleagues, (xi) social responsibility, (xii) nondiscrimination, (xiii) competence, (xiv) legality, (xv) animal care and (xvi) human subject protection [7]. Unfortunately, such strong and mandatory authority is unavailable in case of research publications.

1.1. Aims

This review is undertaken to discuss how medical publications might have abused various ethical norms not only while conducting research but also during the publication process. The review also discusses the possible preventive measures against unethical practices of research and publications.

1.2. Manifestations of medical research misconduct

There are several ways in which ethical violation in medical research are noticed, namely altering instrumentation or research procedure, nonreplicable findings, copying ideas, copying results, false study design, inadequate data, falsifying ethical consent, image manipulations, plagiarism, duplicate publication, etc. These unethical practices are taken place at each of the steps in research, that is performing works to disseminating knowledge through scientific publications [8, 9]. The most common research misconduct, which is manifested through publications, is falsifications and fabrication of data. As per NIH, 'fabrication' means 'the intentional act of making up data or results and recording or reporting them' whereas 'falsification' is manipulating research materials, equipment or processes, or changing or omitting/suppressing data or results without scientific or statistical justification, such that the research is not accurately represented in the research record [10]. It is noticed that 'figures' and 'graphics' where maximum fabrications or falsification take place. The graphical manipulations are mainly through Photoshop and journal editors are struggling hard to fight against these hi-tech manipulations on research data [11, 12]. Overall, we can say that research misconduct manifestation is multidimensional. These may be classified as (1) General research misconduct, (2) Research application misconduct, (3) Data generation misconduct, (4) Financial misconduct, (5) Behavioural misconduct and (6) Publication misconduct.

The foremost important manifestation is *general research misconduct*, which includes fabrication, falsification and plagiarism. These three unethical research practices are very serious offences as it makes research either misrepresented by the facts or underrepresented by the truth. Usually, such unethical practices in research are due to peer pressure and personal gains and pressure from research sponsors. *Research application misconduct* usually occurs while adopting wrong or poor research design or technical errors during experimental, computational and statistical analysis. Improper uses of human subjects, patients or animals also lead to research misconduct and result in ethical violations. *Data generation misconduct* includes false data generation, not including real data in research, not sharing true data with colleagues' especially multicentre studies. *Financial misconduct* in research usually includes misuse of research funds like unauthorized purchase procedures, use of research funds for personal reasons, disclosure of conflict of interest, etc. *Behavioural misconduct* covers inappropriate behaviours towards colleagues, research scholars and gender and religious insensitivities on students, colleagues, patients and subjects. Usually, *publication misconduct* occurs due to authorship dispute, ghost and gifted authorship, plagiarism, duplicate publication and suspicious clinical trials. Study on misconduct in clinical trials found that the most serious forms of research misconduct in clinical trials are selective and biased reporting [12, 13].

1.3. Factors that influence research misconduct

Various factors actually induce research misconduct like:

1. Publish or Perish pressure.
2. Severe competition for funds.
3. Promotion or career advancement policies.
4. Pressure from research sponsors to obtain desired results.
5. Lack of knowledge on research ethics.
6. Desire to 'go ahead'.
7. Personal characters.

In most of the cases, research misconduct is suspected, identified and reported by colleagues. Usually, researchers who work alone and never allow others to observe his or her research

http://dx.doi.org/10.5772/64947

works or researchers who are self centric and do not have an attitude to work in a team are primarily prone to do research misconducts. Research findings in medical sciences should be always repetitive at any place and anytime. Failure to repeat research results by one's own laboratory or external laboratories definitely suspect misconduct.

1.4. Questionable research practices in medicine

These are some criteria which are not direct research misconduct but definitely raise suspicion:

1. Failing to retain significant research data for a reasonable period.
2. Maintaining inadequate research records, especially for results that are published or are relied on by others.
3. Conferring or requesting authorship on the basis of a specialized service or contribution that is not significantly related to the research reported in the paper.
4. Refusing to give peers reasonable access to unique research materials or data that could support published papers.
5. Using inappropriate statistical or other methods of measurement to enhance the significance of research findings.
6. Inadequately supervising research subordinates or exploiting them.
7. Misrepresenting speculations as fact or releasing preliminary research results, especially in the public media, without providing sufficient data to allow peers to judge the validity of the results or to reproduce the experiments.

1.5. Reasons for questionable research practices

The reasons for questionable research practices may be due to poor supervision, excessive workloads, poor training in research, lack of interest of researchers and being over ambitious. These can be found in principal investigators, study coordinators, research scholars, administrative staff, technicians and even the research subjects themselves.

1.6. How questionable research misconduct is done?

The following are some examples of questionable research misconduct [14]:

- Dates misrepresented.
- Duplicate X-rays: different names.
- Blank laboratory reports to fill in.
- Fake subjects: obituary names.
- Analysis done after subjects died.
- Same subject different names.

- Nonexistent subjects created.
- Dates changed in records to match washout periods.
- Consent not signed before entering the study.
- Unqualified staff doing research.
- Inadequate records.
- Failure to get IEC/IRB approval.
- Failure to report changes in research.
- Bogus laboratory results reported.
- Samples study from only a few subjects.
- Subjects received prohibited medication while on study.
- Failure to report adverse events.

Hence, we can say that from knowledge generation (ethical research) to knowledge dissemination (ethical practices and publications) – medical ethics is a common component of research integrity and medical science research cannot afford to lose this integrity for the interest of the humanity (**Figure 1**).

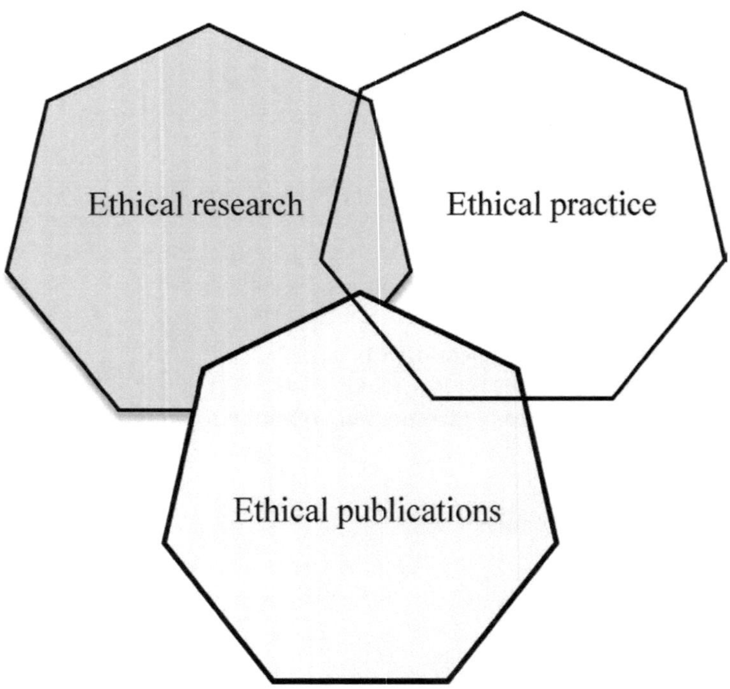

Figure 1. Research integrity through ethical research, practice and publication.

http://dx.doi.org/10.5772/64947

2. Publication ethics

Graf et al. [15] said that academic publishing depends mainly on 'trust'. In the system of research publication procedure, editor trusts reviewers, authors trust editors by expecting fair reviewing processes and finally readers trust authors, reviewers and editors for providing honest sciences. In general, common public outside of research community considers physicians and scientists are just demigod with high morale and integrity. 'Scientists are generally perceived as well-intentioned seekers of truth; universities, as cathedrals of learning and as producers of knowledge vital to the health and welfare of society' [16]. Unfortunately, reports of unethical research publications shake the public confidence on medical scientists. Although medical practitioners, teachers and researchers can recognize publication misconduct and ignore that to some extent, chances of un detection of mistakes and doubtful observations are also may lead to serious consequences. Thorough understanding of publication ethics in medical research is need of the hour. World Association of Medical Editors (WAME), International Committee of Medical Journal Editors (ICMJE) and Committee on Publication Ethics (COPE) are the guiding force to interpret ethical publication appropriately [14]. They have provided guidelines on the publication ethics policies for medical journals on various issues such as study design, authorship, peer review, editorial decisions and plagiarism and also further guided the procedural guidelines to tackle those publication misconduct. These bodies also enlighten editors on various issues such as conflict of interest, authorship disputes, redundant publications, fabrication of data, plagiarism and human and animal rights [17].

2.1. Ethical issues

2.1.1. Why do publication ethics matter?

Published research influences other researchers and establishes credibility for individual or journal. Honest scientific reports build trust among peers and within scientific community. Publication ethics is not confined to one country—it is global by approach and is commonly held throughout the world. Author's seven deadly sin: **Table 1** depicts unethical practices of authors.

2.1.2. Plagiarism

In the era of copy and paste, an excessive dependence on search engine make plagiarism a universally popular among the medical scientists who like to prefer a short cut for the way of success in publication. Plagiarism is defined as 'to copy ideas and passages of text from someone else's work and use them as if they were one's own' [18]. The word plagiarism may be further extended to unreferenced use of the ideas of others submitted as a 'new' paper by a different author! One must know plagiarism may not be considered always as accidental. The most vulnerable part for plagiarism in any research publication is 'methods'.

Another form of plagiarism is self-plagiarism where author copy and paste from his/her previous publications including results, tables and figures without providing copyright clearance certificate from publishers. Self-plagiarism is also an equal crime or research misconduct like simple plagiarism. Fortunately, due to the availability of many anti-plagiarism softwares, this menace has cut down notably. Editor must make his/her peer reviewers alert and possibly train them on this issue. Universities, medical institutes and funding authorities should also sensitize its medical researchers and practitioners on it. The best way to avoid plagiarism is to cite other's work always in the research articles, put the cited words in quotation marks and seek permission from appropriate authorities for references to cite tables, figures, etc. COPE has given a very useful guideline through flow chart on plagiarism for both submitted manuscripts and published manuscripts [18]. The Committee on Publication Ethics (COPE) is a UK-based charitable organization (established 1997) working mainly on research and publication misconduct. COPE has provided some very authentic guidelines addressing publication ethics from authors, reviewers, editors and publishers' point of views. COPE defines the good practices in publication of research articles, which are really very helpful for authors, readers, editors, peer reviewers, editorial board members and journal and book publishers. COPE is the first organization, which advocates accountability of research institutions for its employee scientist's misdeed [19]. ICMJE also directs authors and editors to follow COPE guidelines in case of suspected unethical practices on publications or any ethical dispute [20].

Sl. no.	Sin	Example
1	Carelessness	Citation bias, understatement, negligence
2	Redundant publication	Same tables or literature review reported without noting prior source
3	Undeclared conflict of interest	Failure to cite funding source
4	Unfair authorship	Failure to include eligible authors, honorary authors
5	Human/animal subjects violations	No approval from review board or ethics committee
6	Plagiarism	Reproducing others' work or ideas without as one's own
7	Other fraudulences	Fabrication or falsification of data, misappropriation of other ideas or plans given in confidence

Table 1. Author's ethical misconduct.

2.1.3. Redundant publications

Redundant or duplicate publication is another serious issue pertaining to ethical publications. It is often revealed by reviewers and readers. In the modern era of Internet, it is relatively easy to find out such unethical publication in the form of duplicate publication. Many times, it happens without the knowledge of co-authors or the group of researchers who published it in previous journal. This unethical publication actually causes serious damage on humanity [21]. It makes waste of time of peer reviewers and editors, waste journal print pages unnecessarily. Redundant publication sometimes assault on academic reward system. It also violates

http://dx.doi.org/10.5772/64947

copyright acts and inflated number of publications that injure society as a whole. Below, the facts that make redundant or duplicate publications are mentioned:

- Data in conference abstract? – **No**
- Same data, different journal? – **Yes**
- Data on website? – **May be**
- Data included in review article? – **No if permission is taken**
- Expansion of published data set? – **Yes**

There are certain norms that may help clarify further on duplicate publications like if one takes an approval from both the journals and subsequently publishes, it may not be considered as 'duplicate'. Secondary version for paper intended to different language readers with appropriate permission may not be taken as 'duplicate'. But in any case secondary version faithfully reflects data and interpretations of the primary version with a clear message that it is the secondary publication for journal 'y' based on previously published article (primary) in journal 'x' in 'z' language. COPE has given some useful guidelines on how to handle suspected redundant (duplicate) publications, especially for journal editors. COPE instructed that at the beginning editor must verify whether it is the case of major or minor redundancy. Major redundancy is always considered with evidence of deliberate duplication such as changes of title and data sheet with identical findings. Minor redundancy is something 'salami publication' types with looks of extended follow-up of previously published article. Whatever it may be, editor must contact corresponding author and ask explanation, if satisfied, do not take any action. If it is not found satisfactory, editor has many choices such as inform the incidence to author's superior organizational authority/employer or prompt rejection of manuscript or notice of retraction immediately [22].

2.1.4. *Authorship disputes and ethical misconducts*

Probably, one of the most discussed and complex ethical violation in publication in medicine is authorship disputes and ethical misconducts. The difference between 'disputes' and 'misconduct' may be proclaimed as follows:

Disputes—'Question of interpretation' like whether 'contribution' by the authors was substantial? Whether authorship criteria were discussed when research was planned ? Or it was decided before submission of manuscript?

Misconduct—Authorship is unethical like 'gift' or 'ghost' authorship.

Regarding authorship issue, International Committee of Medical Journal Editors (ICMJE) guidelines states 'anyone who has made a substantial contribution to the conception, design or acquisition of data or analysis and interpretation of data, drafting or revising the article for intellectual contents, or participated in the final approval of the version to be published is entitled to be an author' [23]. The studies revealed that 'gift authorship' is prevalent among authorship misconducts. Gift authorship is usually taken place when research or administrative hierarchy comes in to the picture or because of a colleague with whom we have a personal relationship like son/daughter or husband/wife/relatives etc [24]. But senior researchers or administrative boss who have substantial contribution on the subject at any point like writing manuscript, editing manuscript, reviewing manuscript and providing additional knowledge with high intellectual input on writing science are not considered as 'gifted'. One must clearly remember that simply helping research by way of logistic supports such as sample collections, patients supply, chemicals and reagents supply, helping data collections or providing research funds are not the criteria to become an author [1, 25]. Another unethical authorship dispute is 'ghost authorship'. Ghost authors are the researchers who writes the research article without acknowledgement. This is very common for many cases where researcher drafts an article at the behest of pharmaceutical company. Here, the real author's name never comes in domain of publication. The problem of the ghost author is that whatever they write may not always be correct interpretation and may be biased; hence, it badly affects the researcher community. COPE, ICMJE and WAME have given certain guidelines to tackle this publication misconduct issue. Following are the summarized form of guidelines:

- Journals must have clear authorship criteria.
- Authors should disclose all contributors, regardless of author status and their specific individual contributions and affiliations.
- Authors must sign about their contributions details.
- Authors should disclose any of his/her conflict of interest and a statement whether they have received any support from medical writers [26].

Hence to be précise, it may be stated that as per ICMJE guidelines, the three important mechanics of authorship are 'intellectual input in research, contribution in writing and final written approval of the manuscript' [23, 27]. ICMJE also specifies that authorship criteria should be based only on:

- Substantial contributions to conception and design, or acquisition of data, or analysis and interpretation of data.
- Drafting the article or revising it critically for important intellectual content.

Examples of publication misconduct are authorship disputes and misconducts, which are very common in the medical professionals. Various studies in this regard showed the nature and execution of such unethical practices among medical professional. Works of Dhingra and Mishra [3] revealed that majority of respondent on questionnaires confirm publication misconduct especially authorship disputes among Indian biomedical researchers. Another study of Das et al. [25] observed clear authorship misconduct among medical faculty members of India. In their study, they have found that around 81.4% respondents from medical faculty members confessed authorship disputes in any form among themselves. Further, a comparative study with pharmacy faculty members the dispute level was found to be 29.2%, which further indicates that medical researchers are more vulnerable to authorship misconduct. The study also showed that 74.07% of medical and 68.29% of pharmacy faculty members did not have any discussion on authorship issues at any time before they actually started drafting article for publication. About 88.88% of medical and 36.5% of pharmacy faculty members also mentioned that their professors and head of the departments were included as author although they do not have any contributions or they do not fit in ICMJE authorship criteria. About 81.4% of medical and 29.26% of pharmacy faculty members also mentioned in questionnaires that senior research colleagues interfered while writing manuscript to include their names in the drafted manuscripts. Das et al. further elaborated that even though pharmacy faculty members are better practioners of ethical authorship as compared to their medical counterparts still more sensitization is needed for them to realize ethical authorship [25]. To regulate authorship disputes and misconduct the role of corresponding author should be considered the most important one although other co-authors are also accountable. Every author must have substantial research contribution to justify their inclusion as author. All authors must take their responsibility on manuscript's every pros and cons. The accuracy of all the data, conflict of interest, disclosure of funding authority and get manuscript checked by all the co-authors are the responsibilities must be put on corresponding or principal author's shoulder [9].

2.1.5. Conflict of interest

One of the important but less admitted examples of publication misconduct is nondisclosure of conflict of interest. It may be financial (industry sponsor research) or others like personal interest like employment interest, promotion or career advancement interest, patents, personal believes, grant providing, relationship, academic competition or intellectual passion, etc [23, 28]. Most of the journals make disclosure of potential conflict of interest mandatory and do not publish articles even after acceptance if COI is not disclosed. It has been reported in a study on five leading medical journals like Annals of *Internal Medicine, BMJ, Lancet, JAMA, New England Journal of Medicine* that only 52 of total 3642 articles disclosed their potential COI, that is only 1.4% of total [29]. Another study also showed that one in three lead authors had financial interests in their research by patents, shares or payments for being on advisory boards or as a director, etc [30]. A study conducted by Das et al. [31] on awareness of COI among medical scientists/researchers from India showed that only 12% authors understand COI issues correctly and 19 % of medical authors just heard about it. Very interestingly, the authors who had clear knowledge on COI confessed that hardly they provide COI statement to the journal. The study also found that knowledge of COI is equally poor even among peer reviewers (30%) and editorial board members (25%) too! Some peer reviewers even stated that they are biased toward articles submitted by their known colleagues from medical sciences [31]. Another study also showed that there are no clear guidelines for institution and industries are other cause of COI-related issues [32]. In the complex scenario of COI issues among medical publications, editor of the journals, peer reviewers, research institutions or universities and grant providers must pay more attention to tackle this unethical issue in publications [32]. Das suggested COI case comes out even after publication, in which, the publisher and editor may apologize and issue a formal correction and subsequently retract the article [33].

2.1.6. Fabrication and falsification of data

Fabrication means cooking up data or results (fictitious by nature) as per the hypothesis of research and publishes it in a journal whereas falsification is simply manipulating data or results. It also includes figures or graph distortions. Fabrication also covers selective reporting where authors just report a small number of significant values of the study but hide large number of insignificant observations. Such biasness completely destroys the spirit of science. Normally, both fabrication and falsification of research observations are common for clinical trials (pressure from sponsors) and research activities of medical researchers who have a tendency to go alone instead of working as a research team [1]. In an interest meta-analysis study, Fanelli [9] reported that around 2% of studied medical scientists confessed that they had fabricated or falsified research data. Nearly, one-third of the said study group also confessed that they allowed many publication malpractices including 'dropping data' results of a study in response to pressures from a funding source [9]. The issues on fabrication and falsification of data are very serious by nature, and unfortunately, even the world's top medical research institute faculties are also involved in it. Story of John Long, a pathologist at Harvard Medical School was compelled to resign after publication of his false and fake results on molecular immune complexes related to Hodgkin's disease [34]. Similarly, one Dr. Vijay Soman of Yale University was found an offender on publication ethics because of fabrication and falsification of data from his colleague [35]. This problem is not only restricted on medical researcher/ author, but it is even extended to editor also. Malcolm Pearce who was an Assistant Editor of *British Journal of Obstetrics and Gynaecology* was found to be a publication ethics defaulter. His false case report based on a patient who had gone under successful delivery after reimplanting an ectopic pregnancy was actually nonexisting. Later, his all papers were retracted from various journals [36]. One of the classic example of data fabrication is the story of Ram B. Singh between India. Dr. Singh submitted nine papers from 1992 to 1996 on his research on diet and myocardial infarction. The then Editor of BMJ Professor Richard Smith suspected on Dr. Singh's work and asked him to produce raw data. Dr. R. B. Singh failed to produce that and insisted that data were 'eaten by termites'. It was also found that the institution where he did his research was owned by his family members. BMJ initiated an independent inquiry and published his story [37].

3. Publication ethics

3.1. Best practices

Based on ICMJE and COPE guidelines for publishers, editors, peer reviewers and authors must practice and train themselves against publication misconduct. One of the most important things to promote ethical publication is to encourage research integrity among medical researchers. COPE advocated for a research integrity officer in each of the research institution to monitor and guide various issues pertaining to research ethics including publication ethics [18]. Research Institutions share a responsibility with all of its researchers to preserve scientific integrity in research. They bear the primary responsibility for promoting a culture of good scientific conduct among researchers and students and for the prevention, investigation and punishment of scientific misconduct in their midst. One must remember that research integrity requires the highest professional standards by a critical, open-minded approach, frankness and fairness with absolute honesty.

http://dx.doi.org/10.5772/64947

3.1.1. Publishers, editors and peer reviewers

An editor must take into consideration some important points before sending manuscript for reviewing like whether competing interests are cited by authors or reviewers, ensures that reviewer is adequately qualified and can keep confidentiality and also protects the whistle blower in case of reports on publication misconduct. It is suggested that journal editors must provide a link to WAME or COPE or ICMJE for authors, readers or reviewers to get first-hand information on ethics in publication. Editors should encourage peer reviewers to consider ethical issues on research manuscripts while reviewing and may also ask additional information from authors if need arises. Journals editors and publishers must protect confidentiality of research that includes identity of subjects/patients, etc besides identity of reviewers. Editors may also verify institutional review board clearance on each of the research manuscript in medical journal [15]. Ethical publication also includes timely peer reviewing and publication of the manuscript which is the responsibility of editor and publisher. Authors' especially medical authors always should be sensitized by editors, publishers and institutions that medicine is a profession based on 'absolute trust, philanthropy and altruism'. For ethical publication, the great role of peer reviewers must also to be remembered. Reviewer should be competent enough to review the content of manuscript; he/she should not be in hurry, no COI issues, have knowledge on publication ethics. One more important point on best practice for editors is to remain cultural and gender sensitive on any article. They should carefully observe whether any cultural offence is in the content of manuscripts. Language of the authors should not offend anyone among the readers [15].

3.1.2. Prevention

To regulate appropriately on the issues of ethical publications, institutions or universities should be accountable by the journal publishers for any unethical publication practices authored by the researchers belong to that institution. COPE or ICMJE have given some guidelines but that do not make institutions of author as accountable for any publication misconduct. Institution must have clear and transparent functioning on not only ethical research policy but also on ethical publications. Institution of authors and journal must take a special attention on the clinical trial-based publications. A Strong peer-reviewing system, uses of latest technological support, strong publication ethics policies, active monitoring, protection of whistle blowers and more liaisons between journals and research institutions or universities possibly prevent publication misconduct effectively.

In a summary, we may say that the following points may be considered to prevent publication misconduct:

- Better education on publication guidelines and ethics.
- Introduction of registers for planned and ongoing clinical trials.
- Change criteria from quantity to quality when papers are used for assessment of posts or grants.
- Punish the culprits but be careful that innocent is not victimized.

Acknowledgements

The authors thank Professor B.G. Mulimani, Chief Adviser and former Vice Chancellor of BLDE University, Vijayapura, India, for his encouragement on ethical practices in all the aspects of life in BLDE University.

Author details

Kusal K. Das[1*] and Mallanagoud S. Biradar[2]

*Address all correspondence to: kusaldas@yahoo.com; research@bldeuniversity.ac.in

1 Laboratory of Vascular Physiology and Medicine, Department of Physiology, BLDE University's Shri B. M. Patil Medical College, Hospital and Research Centre, Vijayapura, Karnataka, India

2 Department of Medicine, BLDE University's Shri B. M. Patil Medical College, Hospital and Research Centre, Vijayapura, Karnataka, India

References

[1] Jain AK. Ethical issues in scientific publication. Indian J Ortho. 2010;44(3):235–237. doi: 10.4103/0019-5413.65133

[2] Benos DJ, Fabers J, Farmer J, Jessica P, Gutierrez, Hennessy K, et al. Ethics and scientific publication. Adv Physiol Educ. 2005;29:9–74.

[3] Dhingra D, Mishra D. Publication misconduct among medical professionals in India. Indian J Med Ethics. 2014;11(2):104–106.

[4] Bornmann L, Nast I, Daniel, H-D. Do editors and referees look for signs of scientific misconduct when reviewing manuscripts? A quantitative content analysis of studies that examined review criteria and reasons for accepting and rejecting manuscripts for publication. Scientometrics. 2008:77:415–432.

[5] Fanelli, D. Redefine misconduct as distorted reporting. Nature. 2013;494:149–149.

[6] American Psychological Association. Ethical principles of psychologists and code of conduct. American Psychologist. 2002;57(12):1060–1073.

[7] Resnik DB, Master Z. Policies and initiatives aimed at addressing research misconduct in high-income countries. PLoS Med. 2013;10(3):e1001406.

[8] Mayntz, R. Wissenschaftliches Fehlverhalten: Formen, Faktoren und Unterschiede zwischen Wissens chaftsgebieten. In Ethos der Forschung - Ethics of Research; Max-

Planck-Gesellschaft, Ed.; Max-Planck-Gesellschaft: München, Germany, 1999; pp. 57–72.

[9] Fanelli D. How many scientists fabricate and falsify research? A systematic review and meta-analysis of survey data. PLoS One. 2009;4(5):e5738. doi:10.1371/journal.pone.0005738

[10] Gilbert FJ, Denison AR. Research misconduct. Clin Radiol. 2003;58:499–504.

[11] Balaram P. Plagiarism: a spreading infection. Curr Sci. 2005;88:1353–1354.

[12] Al-Marzouki S, Roberts I, Marshall T, Evans S. The effect of scientific misconduct on the results of clinical trials: a Delphi survey. Contemp Clin Trials. 2005;26:331–33

[13] de Vries R, Anderson, MS, Martinson, BC. Normal misbehavior: scientists talk about the ethics of research. J Empir Res Hum Res Ethics. 2006;1:43–50.

[14] Wager E. Ethical publishing: the innocent author's guide to avoiding misconduct. Menopause Int. 2007;13:98–102

[15] Graf C, Wager E, Bowman A, Fiack S, Scott-Lichter D, Robinson A. Best practice guidelines on publication ethics: a publisher's perspective. Int J Clin Pract. 2007;61(Suppl. 152):1–26.

[16] La Follette, MC. The evolution of the "Scientific Misconduct" issue: an historical overview. Proc Soc Exp Biol Med. 2000;224:211–215.

[17] Abraham P. Duplicate and salami publications. J Postgrad Med. 2000;46:67.

[18] Committee on Publication Ethics. What to do if you suspect plagiarism (a) Suspected plagiarism in a submitted manuscript: Available at http://publicationethics.org/files/Suspected%20plagiarism%20in%20a%20submitted%20manuscript%20%281%29.pdf. Accessed on 29/06/2016.

[19] Wager E, Kleinert S. Cooperation between research institutions and journals on research integrity cases: guidance from the committee on publication ethics. Saudi J Anaesth. 2012;6(2):155–160. doi:10.4103/1658-354X.97030

[20] International Committee of Medical Journal Editors (ICMJE). Available from: http://icmje.org/author. Accessed on 04/06/12016.

[21] Lacetera N, Zirulia L. The economics of scientific misconduct. J Law Econ. Organ 2011;27:568–603.

[22] Committee on Publication Ethics. What to do if you suspect redundant (duplicate) publication? Available at: http://publicationethics.org/files/u2/01B_Redundant_Published.pdf. Accessed on 30/06/2016.

[23] ICMJE, Uniform Requirements for Manuscripts Submitted to Biomedical Journals: Ethical Considerations in the Conduct and Reporting of Research: Authorship and Contributorship. http://www.icmje.org/ethical_1author.html. Retrieved on 17/05/2016.

[24] Matías-Guiu J, García-Ramos R. Fraud and misconduct in scientific publications. Neurologia 2010 Jan–Feb;25(1):1–4. Spanish.

[25] Das KK, Vallabha T, Patil SM, Ray J, Deepak KK. Knowledge on ethical authorship: a comparative study between medical and pharmacy faculty. J Young Pharm. 2016;8(2): 136–140. doi:10.5530/jyp.2016.2.15

[26] Gøtzsche PC, Kassirer JP, Woolley KL, Wager E, Jacobs A, Gertel A, et al. What should be done to tackle ghostwriting in the medical literature? PLoS Med. 2009;6(2):e1000023. doi:10.1371/journal.pmed.1000023

[27] Al Lamki L. Ethics in scientific publication: plagiarism and other scientific misconduct. Oman Med J. 2013;28(6):379–381. doi:10.5001/omj.2013.112

[28] FACTSHEET Conflict of Interest. Elsevier | Ethics in Research & Publication Available at: https://www.elsevier.com/__data/assets/pdf_file/0010/92476/ETHICS_COI02.pdf. Accessed on 13/05/2016.

[29] Hussain A, Smith R. Declaring financial competing interests: survey of five general medical journals. BMJ. 2001;323:263–4.

[30] Bekelman JE, Li Y, Gross CP. Scope and impact of financial conflicts of interest in biomedical research. A systematic review. JAMA. 2003;289:454–65.

[31] Das KK, Vallabha T, Ray J, Murthy PSN. Conflict of interest – serious issue on publication ethics for Indian Medical Journals. J Nepal Med Assoc. 2013;52(190):357–60.

[32] Lexchin J, Bero LA, Djulbegovic B, Clark O. Pharmaceutical industry sponsorship and research outcome and quality: systematic review. BMJ. 2003;26:1167–70.

[33] Das KK. The editorial policy on conflict of interest – a major issue for the Indian Medical Journals. J Krishna Inst Med Sci. 2012;1(1):4–5.

[34] Hamblin TJ. Fake. Br Med J (Clin Res Ed). 1981;283:1671–4.

[35] Altman L, Melcher L. Fraud in science. Br Med J (Clin Res Ed). 1983;286:2003–6.

[36] Smith R. Research misconduct: the poisoning of the well. J R Soc Med. 2006;99:232–7.

[37] Smith J, Godlee F. Investigating allegations of scientific misconduct. BMJ. 2005;331:245–6.

Chapter 4

Physicians and Pharmaceutical Industry: Need for Transparency by Conflict of Interest Declaration and Independent Ethical Oversight

Frieder Keller, Krzysztof Marczewski and
Drasko Pavlovic

Additional information is available at the end of the chapter

http://dx.doi.org/10.5772/65104

Abstract

Aim: The collaboration between physicians and pharmaceutical industry are based on financial interests on both sides. Transparency will bring the scientific as well as social public to a position from which they are able to judge whether the physician's interest dominates over the patients' benefit.

Proposals: The declaration of any conflicting interest (CoI) must be placed on the first slide of a scientific presentation and on the last line of a published abstract. Declaring the amount of industry funding in the patient information form of all investigator-initiated clinical trials should also be considered. To limit influence of the industry on data presentation and interpretation, employees or experts acting in charge of a sponsoring company cannot be made co-authors but will only be mentioned in the acknowledgement of publications. For approval of a clinical trial the local ethics committee or the institutional review board should evaluate whether industrial funding is balanced by personal work.

Conclusion: The need to disclose will motivate physicians to keep their own interests under control. The intrinsic motivation of the profession needs support by external oversight to maintain the ethics in physician-industry interactions. (Expenses were refunded for FK, KM and DP by ERA-EDTA.)

Keywords: ethics, disclosure, conflict of interest, funding, physician-industry relationship

1. Introduction, background and objectives

Pharmaceutical funding plays an important role in medical progress. Thus, clinical and academic research has been significantly commercialized [1]. There are data to suggest that economic interest from industry may have a negative influence on the objectivity of the science, the publication of research, and even patient management as discussed for rofecoxib [2, 3]. The manufacturing company for instance had engaged "… in misleading practices to promote the prescription and usage of rofecoxib, including 'fake' journals and guidelines to 'drug reps' that minimised the adverse cardiovascular risks" [4]. Industry-sponsored reports are up to four times as likely to favour a pharmaceutical company's product compared to independently published data [5, 6].

Influence of industry interest is not always zero as in the case of the British National Institute of Clinical Excellence (NICE) where cooperation with industry is prohibited for all members. But influence can be suspected in 50–100% of other expert panels [3, 7, 8]. Such influence has been found in guidelines promoting, for instance, gabapentin or preferably recommending patented antidiabetic drugs [7, 9]. Pushing epoetins and cinacalcet, KDOQI guidelines had favoured a target haemoglobin up to 12 mg/dl and recommended calcium values less than 9.5 mg/dl (<2.38 mmol/l) against other evidence [10].

We wish to analyse the need for clear regulations and we will make new proposals for how best to disclose or – less adversarial – declare a conflict of interest (CoI). On the one side and as a consequence, it has been suggested that the dynamics of this process need to be more restricted and governed by less tolerant regulation [11]. Considerations of how ethics here could be made effective might help from another side.

2. Analysis of the present state

Regulations such as the "Physician Payment Sunshine Act" are needed but probably not sufficient to make sure that conflicts are declared and not concealed [12]. Even such measures as the proposed Center for Monitoring and Implementing Publication Ethics will only be instrumental when clear sanctions are laid down [13]. But the consequences regarding sanctioning of malpractice are unclear. While misuse and non-adherence call for legal regulations or even punishment, responsible and trustworthy actions have a foundation in ethics. Physicians must have an intrinsic interest that patients and colleagues trust what he or she is saying and doing.

In medicine generally, "… the primary interests are the health of the patient" whereas financial gain, prestige or preferences are not illegitimate but secondary interests [14]. Medical professionalism "… demands placing the interests of patients above those of the physician" [15]. Such privilege likewise no client could expect from an investment banker. According to the charter of medical professionalism, the primary patients' interests are: welfare, respect for autonomy and social justice – namely the main principles of medical ethics [15, 16]. The primacy of patient welfare can be guaranteed only if the indication for treatment is medical, not economical. The patient's autonomy will be compromised if she/he was misled and thus consented to a therapy that appeared better than what it ended up being. Financial interests can do harm to the principles of parsimony and social justice when the more expensive intervention is not justified by evidence but favoured by a physician under pharmaceutical influence.

Economic mechanisms benefitting the market need other and special regulations in medicine in which the person who decides, benefits and pays is not one and the same. Market mechanisms do not work in medicine where the interests are not equivalent and health cannot be circulated like money, exchanged like a ware or consumed like goods. To give an example, dialysis will be mentioned here. Financial incentives will motivate adequate measures to reduce mortality and facilitate access to dialysis for all patients who need it. But such economic interest could also corrupt the physician to prematurely recruit patients for dialysis or unnecessarily maintain this treatment [17].

2.1. Need for specified regulations

Whether the interests of professionals really are secondary to the welfare of the patient must be made clear and transparent to all persons concerned. As professionals, physicians must seek patients' confidence, social trust and vocational reputation [3]. In the patient-physician relationship, ethics is the foundation of confidence, and confidence is the foundation of sustainability.

Although the percentage of physicians with any relationship to industry decreased, this relationship was still very high at 84%, ranging from a $10 sandwich to the $1 million research grant [3, 18, 19]. Expectations of reciprocity may be the primary reason for sponsorship by industry [20].

Due to subtle psychological effects, even small gifts might be a powerful stimulus for reciprocity [21]. Physician who received a single industry-sponsored meal had significantly higher rates of prescribing rosuvastatin, nebivolol or desvenlafaxine [19]. Research into the psychology of receiving and giving gifts indicates that more appropriate regulations would be necessary [2, 3]. Most scientists call for an open conflict of interest declaration, although some of them might fail to identify their own personal conflicting financial interest [22]. Whilst 61% of physicians had the opinion that financial incentive did not influence their own practice, only 16% believed that the same was true of their colleagues [23].

It is uniform experience that physicians and scientists do not willingly talk about their own motivations – be they economic or intellectual [21, 24]. The rigor of a study has been judged to be significantly reduced when funding came from industry compared to studies where the funding was from a government agency, for instance the National Institute of Health [25, 26]. Worldly wisdom tells patients and doctors to intuitively distrust studies funded by companies [25].

In response, it has become a prerequisite that all persons involved in the activities of the European Renal Association/European Dialysis and Transplant Association (ERA-EDTA) adhere to the 2014 Council Regulations that were initiated by the ERA-EDTA Ethics Committee (see Acknowledgments): the disclosure – or less adversarial – the declaration of interest (DoI) is mandatory to make transparent whether there could be a conflict of interest (CoI). This applies when presenting a talk or chairing a meeting, when working on guidelines, acting as editor or reviewer and submitting a manuscript for publication. Such declaration of interest regulations are practiced now in most medical societies.

Regulations are needed but probably not sufficient to ensure that conflicting interests are declared and not concealed [12]. Research on the tension between moral rules shows "… that collaborative settings provide fertile ground for the emergence of corruption" [27]. Physicians not always will resist to the seductive influences by industry. Regulations will only be instrumental when clear sanctions are laid down [23].

3. Proposals

Threat of scientific banning, ostracism, litigation and the law should intend to discourage concealment and deception. The German parliament, for example, is planning to issue a new anti-corruption law for all workers in the health system (StGB §299a and b). The boundaries, however, between an illegal incentive that stimulates corruption and a financial reward judged to be an adequate compensation are fluid.

Targets	The economic interests of clinicians and researchers need to be more transparent to: • establish sustainable trust and confidence in clinical science, medical practice and data published from physicians and scientists, • enable patients', audiences' or readers' own judgments as to whether a conflict of interest exists and whether such conflict impacts on science or practice, • discourage economic incentives but limit financial compensation to what has been invested as personal work.
The aim of legal regulations and ethical motivations should be to confine the influence of industry that could impact on clinical practice and on the conduct and reporting of medical research.	

Table 1. Declaration of conflict of interest.

At first, transparency might help better the matter than pending threats to be sanctioned in the case of non-adherence to the canonical conflict of interest regulations. Transparency will help to bring the scientific as well as social public to a position from which they are able to judge whether the physician's interest dominates over the patients' benefit. Fair payment compensates for work performed, while inadequate payment tends to influence decisions with costly consequences.

Proposals	To improve transparency: **1.** The conflict of interest should be declared: **a.** on the first slide of any oral presentation, **b.** on the last line of an abstract as in all NEJM papers (**Figure 1**), and c. at the beginning of official guideline statements. **2.** For the sake of confidence, the dominant conflict of interest should be declared first. **3.** Consider mentioning funding by industry also in the patient information form of any investigator-initiated trial. **4.** Avoid employees of a company influencing the conduct of a trial and bias the presentation of data. It should be part of the primary contract that all contributions by employees or representatives of industry can only be mentioned in the acknowledgement. **5.** Violation of the conflict of interest declaration rules can be efficiently sanctioned when the researcher will no longer be allowed to acquire and receive any further legal funding coming from either industry or from government. To discourage corruption by limiting financial incentives: **1.** The ethics committee must control all clinical research proposals for adequacy of funding and balanced financial compensation. **2.** An ethical oversight should be established for all guideline committees.

Table 2. Proposals for declaration of interest (DoI) whether a potential conflict of interest (CoI) might exist and how to improve transparency.

With the most evident objectives in mind (**Table 1**), therefore, we suggest seven proposals on how to deal with a potential conflicting interest when presenting a talk, when publishing a paper or planning a trial (**Table 2**). More far-reaching standards than for presentations and publications will be needed for committees installed to develop trustworthy clinical practice guidelines as Richard N Shiffman has pointed out (Institute of Medicine, 2011).

3.1. Declaration of conflicts of interest (CoI)

The declaration of interests does not necessarily mean that the interests are in conflict with truth [3]. Conversely, also a declared conflict of interest can still negatively influence science and practice [28]. The subjective feeling of integrity can turn out to be a biased dependency on an objective level [29]. In scientific journals, transparency is needed. The journal editors must judge whether a conflict of interest disclosure is correct, and otherwise reject such work [30]. Not only with regard to financial conflicts but also an intellectual conflict of interest might prevent a fair reviewing process. This can be managed only by the editor [31].

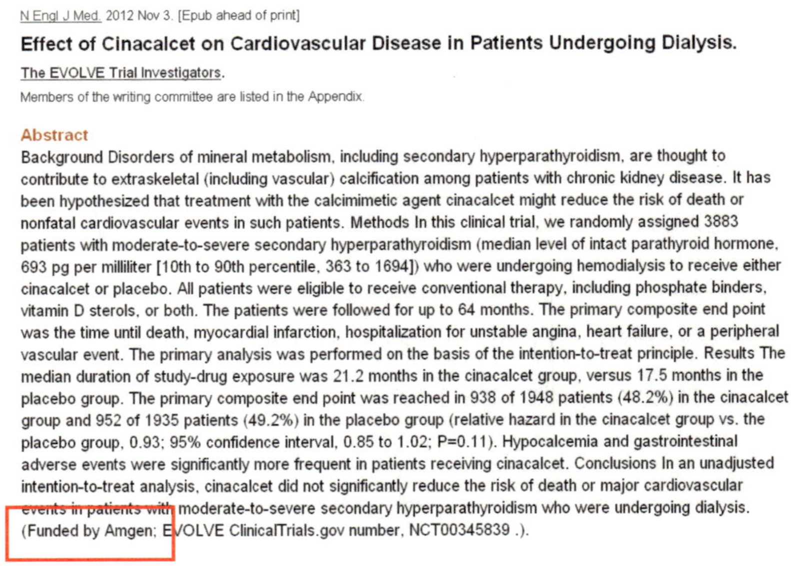

N Engl J Med. 2012 Nov 3. [Epub ahead of print]

Effect of Cinacalcet on Cardiovascular Disease in Patients Undergoing Dialysis.

The EVOLVE Trial Investigators.

Members of the writing committee are listed in the Appendix.

Abstract

Background Disorders of mineral metabolism, including secondary hyperparathyroidism, are thought to contribute to extraskeletal (including vascular) calcification among patients with chronic kidney disease. It has been hypothesized that treatment with the calcimimetic agent cinacalcet might reduce the risk of death or nonfatal cardiovascular events in such patients. Methods In this clinical trial, we randomly assigned 3883 patients with moderate-to-severe secondary hyperparathyroidism (median level of intact parathyroid hormone, 693 pg per milliliter [10th to 90th percentile, 363 to 1694]) who were undergoing hemodialysis to receive either cinacalcet or placebo. All patients were eligible to receive conventional therapy, including phosphate binders, vitamin D sterols, or both. The patients were followed for up to 64 months. The primary composite end point was the time until death, myocardial infarction, hospitalization for unstable angina, heart failure, or a peripheral vascular event. The primary analysis was performed on the basis of the intention-to-treat principle. Results The median duration of study-drug exposure was 21.2 months in the cinacalcet group, versus 17.5 months in the placebo group. The primary composite end point was reached in 938 of 1948 patients (48.2%) in the cinacalcet group and 952 of 1935 patients (49.2%) in the placebo group (relative hazard in the cinacalcet group vs. the placebo group, 0.93; 95% confidence interval, 0.85 to 1.02; P=0.11). Hypocalcemia and gastrointestinal adverse events were significantly more frequent in patients receiving cinacalcet. Conclusions In an unadjusted intention-to-treat analysis, cinacalcet did not significantly reduce the risk of death or major cardiovascular events in patients with moderate-to-severe secondary hyperparathyroidism who were undergoing dialysis. (Funded by Amgen; EVOLVE ClinicalTrials.gov number, NCT00345839 .).

Figure 1. Funding disclosed at the end of the abstract as in all NEJM papers.

The readers and the public must have the opportunity to form their own opinions about the independence and the value of a study [29]. An unstructured list of many sponsors, although unintentional [32], could make a contributor falsely appear to the public to be both prestigious and independent. Any potentially conflicting interest should be declared in an exposed and not a hidden place (**Figure 1**) and a fair and honest interest disclosure states the potentially most important conflict first [1].

The general declaration of interests might bear the risk that a conflict could be perceived that does not exist. But such scepticism will do less harm than the growing mistrust when a real conflict can be concealed. It does not matter here whether this is only a perceived conflict or a real conflict of interest. Physicians and patients do not want to be victims of subtle suggestions [33].

To be informed, not only readers and listeners but also patients need complete transparency. Therefore, it should be considered that the amount of financial funding by the industry must also be stated for each included subject in the patient information form of any investigator-initiated trial (**Table 2**). If not stated for each included patient, the total amount must be declared at least for the complete study. In any case, the disclosure is needed whether the money goes to the institution or into the pockets of the investigator. Mistrust will spread and

http://dx.doi.org/10.5772/65104

neglected sustainability will come back in the end. A damaged reputation ultimately results, as has been discussed with the examples of gene therapy, rofecoxib, new antidiabetics or oseltamivir [3, 34].

Declaration of interest rules are not yet sufficiently specified. Besides being blamed by publicity, one sanctioning instrument might be most efficient: a researcher could be excluded from receiving any further legal sponsoring, be it from industry or from government. But academic institutions dislike inflicting such ultimate and most efficient sanctioning instruments since these harm the institution itself.

3.2. Authorship

To publish purely industry-driven research, the name of a highly recognized scientist can be misrepresented as an author. Ghost, guest or gift authors might make a company look like a great research partner or a paper look like a good science [12]. On the other side, a great scientific and ethical problem is posed when individuals participate in research, data analysis and/or writing of a manuscript but are not named or disclosed in the author byline or acknowledgements [35]. Some ethics experts maintain that placing employees of a company on the list of co-authors will make them accountable and reduce the role of ghost-writers (Susan L Norris, Howard Brody).

If included in the list as co-authors, however, employees or experts acting in charge of pharmaceutical companies can influence the results. As discussed for epoetin or for some psychopharmacological trials, drug company employees could significantly guide the presentation of data [36, 37]. The influence of a co-author could even be more problematic than that of a ghost-author or a ghost-writer. The analysis of 198 trials in three psychiatric journals found that the involvement of a drug company employee was even stronger associated with the interpretation of study outcome than the financial sponsorship [36].

To limit the subtle dominance of companies, the probably better solution could be first to only consult independent, not the companies' statisticians. Furthermore, the contribution to a study by employees or representatives of the industry should exclusively but explicitly be mentioned in the acknowledgements [38]. This does not preclude employees' scientific work from being published independently by themselves. Employees can publish their own papers but should not be made co-authors of investigator-initiated trials. This rule must be clarified by contract from the beginning of the cooperation (**Table 2**).

3.3. Ethical oversight

In consequence, clear guidelines, rules and even laws regarding how to motivate for integrity and to control for misconduct need to be released. We propose to establish both academic and legal measures to meet this target. Also in such research that was publicly funded by a government agency, misleading influences might endanger research integrity: plagiarism, falsification and fabrication of data, fraud, and violation of good scientific practice reflect the other growing debate on misconduct in research and science [39, 40].

Intrinsic virtue prompts the professionals to not let themselves be corrupted [3]. But such virtue should be encouraged and affirmed. The oversight by an ethics committee might refine such virtue. The order to disclose all industry interactions had a salutary effect: transparency facilitated the complete abstinence from the addictive power exerted by industry sponsoring; after enactment of this order all members of the Medicines Agency of the German Physicians Association AkdÄ completely ceased all collaborations with industry [41]. Transparency regulations and oversight ethics have not only a prophylactic but also a curative potential when applied to the physician-industry relationship.

Between the basic confidence in personal virtue of the physicians and the deterrent legal threat by law there is a role for the ethics committees. For all experts with a mandate to release guidelines an ethical oversight has already been postulated [42]. This oversight will demotivate any concealment and manipulations. Such a committee will also be able to correct misconduct and personal failure. This committee will approve whether the financial compensation is in balance with personal work.

Financial incentives will be discouraged by such an institution before the law comes into action. Such an ethics committees and institutional review boards have long been established to approve any clinical trial for the risk-benefit balance. It should be made an obligation to such ethics committees to guarantee also the balance between financial compensation and personal work (**Table 2**).

4. Conclusion

Material goods such as medicinal products or new therapeutic agents can best be manufactured and distributed by market mechanisms; but social and personal relations should still be regulated by moral values [43]. Protection is needed for an independent medical and academic sector where both the patient-physician relationship and integrity of research are "sacrosanct" [1]. True science must not be free from any interest, be it economic or emancipatory [44]. In order, however, to maintain academic integrity and to comply with the fiduciary duty of the medical profession, it is in the interest of the credibility of each scientist and every physician to check herself/himself for possible conflicts of interest and to limit such influence. While patients defend their personal identity, physicians must fight for their moral integrity.

Money will do damage to physicians' reputations especially when it makes them resemble a company lackey [3]. Many physicians feel ashamed when their collaboration with industry is disclosed. Physicians will be motivated to keep their own interests under better control by the need to disclose any CoI and make their interests transparent. A disappointing response to the conflicting interests has been identified as "moral disengagement operations" such as justification, euphemistic labelling, diffusion of responsibility, sharing blame, minimizing risks and victim dehumanization [45].

Where patients can no longer trust medicine, controllers and lawyers will dominate the scene [46]. Transparency consequently allows for more tolerant regulation because the ethical principles are proactive and not restrictive. On the patients' side, in addition, transparency about financial stimuli might also bring down some illusionary hopes and wishes back to reality.

Acknowledgements

The positions in this article are personal opinions of the authors. As former members of the ERA-EDTA Ethics Committee, we discussed the subject during our regular meetings. The two other members of the ERA-EDTA ethics committee were Jacques Bernheim who had the primary idea and Richard Trompeter who substantially contributed to the argumentation with ethical principles. We acknowledge their valuable input. Monica Fontana organized the meetings of the ERA-EDTA Ethics Committee and acted as a moderator in our discussions.

Transparency

The authors declare that they have no competing interests with regard to this article. The authors received travel expenses refunded from the ERA-EDTA. For conflict of interest declarations see http://www.era-edta.org/page-17-79-0-79-ethicscommittee.html. FK had financial connections to Novartis, MSD, Medicine and Aspen Company; KM to Amgen and Roche; DP to Amgen, Pliva, Belupo, Sandoz, Boehringer Ingelheim and Abbvie.

Author details

Frieder Keller[1*], Krzysztof Marczewski[2] and Drasko Pavlovic[3]

*Address all correspondence to: frieder.keller@uni-ulm.de

1 Medical Department I, Nephrology, Ulm University Hospital, Ulm Germany

2 Department of Nephrology, Endocrinology Hypertension and Internal Medicine, Pope John Paul II Regional Hospital, Zamość, Poland

3 Department of Nephrology and Dialysis, Sestre Milosrdnice University Hospital, Zagreb, Croatia

References

[1] Lemmens T. Leopards in the temple: restoring scientific integrity to the commercialized research scene. J Law Med Ethics. 2007; 32: 641-57. Medline: 15807352.

[2] McNeill PM, Kerridge IH, Henry DA, Stokes B, Hill SR, Newby D, Macdonald GJ, Day RO, Maguire J, Henderson KM. Giving and receiving of gifts between pharmaceutical companies and medical specialists in Australia. Intern Med J. 2006; 36: 571-8. Medline: 16911549.

[3] Rosenbaum L. Conflicts of interest: part 1: reconnecting the dots--reinterpreting industry-physician relations. N Engl J Med. 2015; 372: 1860-4. Medline: 25946288.

[4] Faunce T, Townsend R, McEwan A. The Vioxx pharmaceutical scandal: Peterson v Merk Sharpe & Dohme (Aust) Pty Ltd (2010) 184 FCR 1. J Law Med. 2010; 18: 38-49. Medline: 20977160.

[5] Brennan TA, Rothman DJ, Blank L, Blumenthal D, Chimonas SC, Cohen JJ, Goldman J, Kassirer JP, Kimball H, Naughton J, Smelser N. Health industry practices that create conflicts of interest: a policy proposal for academic medical centers. JAMA. 2006; 295: 429-33. Medline: 16434633.

[6] Schott G, Pachl H, Limbach U, Gundert-Remy U, Lieb K, Ludwig WD. The financing of drug trials by pharmaceutical companies and its consequences: part 2: a qualitative, systematic review of the literature on possible influences on authorship, access to trial data, and trial registration and publication. Dtsch Arztebl Int. 2010; 107: 295-301. Medline: 20490338

[7] Norris SL, Holmer HK, Ogden LA, Burda BU, Fu R. Conflicts of interest among authors of clinical practice guidelines for glycemic control in type 2 diabetes mellitus. PLoS One. 2013; 8: e75284. Medline: 24155870.

[8] Schott G, Dünnweber C, Mühlbauer B, Niebling W, Pachl H, Ludwig WD. Does the pharmaceutical industry influence guidelines? Two examples from Germany. Dtsch Arztebl Int. 2013; 110: 575-83. Medline: 24078837.

[9] Kjaergard LL, Als-Nielsen B. Association between competing interests and authors' conclusions: epidemiological study of randomised clinical trials published in the BMJ. BMJ. 2002; 325(7358): 249. Medline: 12153921.

[10] Coyne DW. Influence of industry on renal guideline development. Clin J Am Soc Nephrol. 2007; 2: 3-7; discussion 13-4. Medline: 17699377.

[11] Bindslev JB, Schroll J, Gøtzsche PC, Lundh A. Underreporting of conflicts of interest in clinical practice guidelines: cross sectional study. BMC Med Ethics. 2013; 14: 19. Medline: 23642105.

[12] Ratain MJ. Forecasting unanticipated consequences of "The Sunshine Act": mostly cloudy. J Clin Oncol. 2014; 32: 2293-5. Medline: 24934789.

[13] Ruff K. Scientific journals and conflict of interest disclosure: what progress has been made? Environ Health. 2015; 14: 45-53. Medline: 26024916.

[14] Thompson DF. Understanding financial conflicts of interest. N Engl J Med. 1993; 329: 573-6. Medline: 8336759.

[15] Blank L. Medical Professionalism Project. Medical professionalism in the new millennium: a physicians' charter. Lancet. 2002; 359: 520-2. Medline: 11853819.

[16] Cassel CK, Hood V, Bauer W. A physician charter: the 10th anniversary. Ann Intern Med. 2012; 157: 290-1. Medline: 22910940.

[17] Keller F, Dress H, Mertz A, Marckmann G. Financial incentives and dialysis. Med Klin (Munich). 2007; 102: 659-64. Medline: 17694285.

[18] Campbell EG, Rao SR, DesRoches CM, Iezzoni LI, Vogeli C, Bolcic-Jankovic D, Miralles PD. Physician professionalism and changes in physician-industry relationships from 2004 to 2009. Arch Intern Med. 2010; 170: 1820-6. Erratum in: Arch Intern Med. 2010; 170: 1966. Medline: 21059976.

[19] DeJong C, Aguilar T, Tseng CW, Lin GA, Boscardin WJ, Dudley RA. Pharmaceutical industry-sponsored meals and physician prescribing patterns for medicare beneficiaries. JAMA Intern Med. 2016. doi: 10.1001/jamainternmed.2016.2765. [Epub ahead of print] Medline: 27322350.

[20] Dana J, Loewenstein G. A social science perspective on gifts to physicians from industry. JAMA. 2003; 290: 252-5. Medline: 12851281.

[21] Sah S. Conflicts of interest and your physician: psychological processes that cause unexpected changes in behavior. J Law Med Ethics. 2012; 40: 482-7. Medline: 23061575.

[22] Beran RG. Conflict of interests – criticising the critics. Med Law. 2009; 28: 557-63. Medline: 20157969.

[23] Steinman MA, Shlipak MG, McPhee SJ. Of principles and pens: attitudes and practices of medicine house staff toward pharmaceutical industry promotions. Am J Med. 2001; 110: 551-7. Medline: 11347622.

[24] Kesselheim AS, Wang B, Studdert DM, Avorn J. Conflict of interest reporting by authors involved in promotion of off-label drug use: an analysis of journal disclosures. PLoS Med. 2012; 9: e1001280. Medline: 22899894.

[25] Licurse A, Barber E, Joffe S, Gross C. The impact of disclosing financial ties in research and clinical care: a systematic review. Arch Intern Med. 2010; 170: 675-82. Medline: 20421551.

[26] Kesselheim AS, Robertson CT, Myers JA, Rose SL, Gillet V, Ross KM, Glynn RJ, Joffe S, Avorn J. A randomized study of how physicians interpret research funding disclosures. N Engl J Med. 2012; 367: 1119-27. Medline: 22992075.

[27] Weisel O, Shalvi S. The collaborative roots of corruption. Proc Natl Acad Sci U S A. 2015; 112: 10651-6. Medline: 26261341.

[28] Bes-Rastrollo M, Schulze MB, Ruiz-Canela M, Martinez-Gonzalez MA. Financial conflicts of interest and reporting bias regarding the association between sugar-sweetened beverages and weight gain: a systematic review of systematic reviews. PLoS Med. 2013; 10: e1001578; discussion e1001578. Medline: 24391479.

[29] Taylor PL. Innovation incentives or corrupt conflicts of interest? Moving beyond Jekyll and Hyde in regulating biomedical academic-industry relationships. Yale J Health Policy Law Ethics. 2013; 13: 135-97. Medline: 23815042.

[30] Rifkind AB. Competing interests: expanding rapidly. Nature. 2012; 488: 590. Medline: 22932371.

[31] Roohi E, Mahian O. Some opinions on the review process of research papers destined for publication. Sci Eng Ethics. 2015; 21: 809-12. Medline: 24781873.

[32] Stallone G, Infante B, Grandaliano G, Bristogiannis C, Macarini L, Mezzopane D, Bruno F, Montemurno E, Schirinzi A, Sabbatini M, Pisani A, Tataranni T, Schena FP, Gesualdo L. Rapamycin for treatment of type I autosomal dominant polycystic kidney disease (RAPYD-study): a randomized, controlled study. Nephrol Dial Transplant. 2012; 27: 3560-7. Medline: 22785114.

[33] Tattersall MH, Dimoska A, Gan K. Patients expect transparency in doctors' relationships with the pharmaceutical industry. Med J Aust. 2009; 190: 65-8. Medline: 19236290.

[34] Ross JS, Gross CP, Krumholz HM. Promoting transparency in pharmaceutical industry-sponsored research. Am J Public Health. 2012; 102: 72-80. Medline: 22095335.

[35] Stretton S. Systematic review on the primary and secondary reporting of the prevalence of ghostwriting in the medical literature. BMJ Open. 2014; 4: e004777. Medline: 25023129.

[36] Tungaraza T, Poole R. Influence of drug company authorship and sponsorship on drug trial outcomes. Br J Psychiatry. 2007; 191: 82-3. Medline: 17602130.

[37] Coyne DW. The health-related quality of life was not improved by targeting higher hemoglobin in the Normal Hematocrit Trial. Kidney Int. 2012; 82: 235-41. Medline: 22437411.

[38] Dowsett SA, Van Campen LE, Bednar LA. Developing good scientific publishing practices: one pharmaceutical company's perspective. Curr Med Res Opin. 2010; 26: 1249-54. Medline: 20350148.

[39] Fang FC, Steen RG, Casadevall A. Misconduct accounts for the majority of retracted scientific publications. Proc Natl Acad Sci U S A. 2012; 109: 17028-33. doi: 10.1073/pnas.1212247109. Epub 2012 Oct 1. Erratum in: Proc Natl Acad Sci U S A. 2013; 110: 1137. Medline: 23027971.

[40] Bosch X. Improving biomedical journals' ethical policies: the case of research misconduct. J Med Ethics. 2014; 40: 644-6. Medline 24505117.

[41] Dr Honest and Mr Greed. Physicians must disclose how much money they receive from pharmaceutical industry. DER SPIEGEL, 2016; 28: 98

[42] Schünemann HJ, Al-Ansary LA, Forland F, Kersten S, Komulainen J, Kopp IB, Macbeth F, Phillips SM, Robbins C, van der Wees P, Qaseem A; Board of Trustees of the Guidelines International Network. Guidelines International Network: principles for disclosure of interests and management of conflicts in guidelines. Ann Intern Med. 2015; 163: 548-53. doi: 10.7326/M14-1885. Medline: 26436619.

[43] Sandel MJ. What Money Can't Buy: The Moral Limits of Markets. Allen Line, Pinguine Group, London 2012: pp 131.

[44] Habermas J. Erkenntnis und Interesse. Suhrkamp, Frankfurt am Main 1968: pp 88.

[45] Mecca JT, Gibson C, Giorgini V, Medeiros KE, Mumford MD, Connelly S. Researcher perspectives on conflicts of interest: a qualitative analysis of views from academia. Sci Eng Ethics. 2015; 21: 843-55. Medline: 25115563.

[46] Davis DS, Kodish E. Laws that conflict with the ethics of medicine: what should doctors do? Hastings Cent Rep. 2014; 44: 11-4. Medline: 25412971.

Chapter 5

Bioethics and the Experiences of Hansen's Disease Survivors

Makiko Kondo, Kazuo Mori , Hiroshi Nomura,
Hanako Kadowaki, Makiko Watanabe,
Akemi Doi and Sayaka Shima

Additional information is available at the end of the chapter

http://dx.doi.org/10.5772/65574

Abstract

Historically, Hansen's disease patients suffered from discrimination because their physical features changed due to the bacterium *Mycobacterium leprae* (*M. leprae*) and made them "ugly" in the eyes of society. Former Japanese governments saw them as a national disgrace and forced them to reside in leprosaria. Since the law requiring isolation continued after the silver bullet was developed, survivors could not leave the leprosaria and return to society. Currently, survivors' average age is 82 and they live in 13 national sanatoriums. When they pass away, the history of Hansen's disease in Japan will end, so we must record their experiences. We conducted qualitative and inductive studies with survivors. In this chapter, we reconstruct them from the perspective of bioethics and propose several theories surrounding them: (1) How former leprosaria and medical administrations in Japan threatened bioethical principles; (2) the wisdom of aging survivors, who lived through extreme situations, and what real restoration of their rights might look like; and (3) the ethical dilemmas of how we will care for the survivors—who have multiple severe sequelae—until they all pass away. Finally, we will introduce our ethical nursing practices in relation to caring and understanding via holism.

Keywords: stigma, Hansen's disease, bioethical principles, wisdom, human caring

1. Introduction

For many years, Hansen's disease patients' rights were violated, and the four bioethical principles were ignored during the course of their treatment. In **Table 1**, we show part of the life of Ms. Hanako Kadowaki, who is 84 years old, blind, and a Hansen's disease survivor [1].

Listener: What do you think about the Leprosy Prevention Law?

Mrs. Kadowaki: Because I was child, I did not understand it in detail, but enforced isolation means being treated like a criminal. Although my situation was not severe, the policeman and prefecture administrator came to our village many times. My friend's father was forced into truck to be sent to the leprosarium while he was working. He was carpenter, and he asked the policeman, "Please wait until I have finished building this house." They did not wait and forced him into the truck. This is what enforced isolation is like. I cried when I heard it.

When I entered this leprosarium, it was filled with pilgrims who had been captured. At that time, the emphasis was on catching pilgrims, so this facility overflowed with residents. Pilgrims were gathered at the pier on the mainland. Pilgrims without handicaps were sent to another island, and pilgrims with handicaps were sent to this island. Thus, there were many handicapped pilgrims in this leprosarium at that time.

Because they could not live in their hometowns, pilgrims were working and traveling around on both hot and cold days, and sleeping in front of shrines and temples and begging. They traveled with difficulty, and although they had no medical treatment, they were forced to work, so they became crippled by degrees. After decades of pilgrimage, they entered blind and with missing limbs, and I pitied them. There were lice, there were fleas, and there were bedbugs. It was very terrible, and the pilgrims had nothing except their own bodies, but my isolation was not completely terrible.

At that time, dying patients could not be cared for by the nurse, so they died in a terrible situation. They died weak, without food, saying, "I want to die after drinking myself full of wine," or "I want to eat sweets." I felt pity for those dying people at the time. Their funeral was shabby, too.

The meals were skimpy. When the war was intense, there was no food, no clothing, no soap, and no medical treatment. After the war, there was still little food, and I was hungry, so I asked an outside person to buy food for me, as I was ashamed to ask anyone inside.

At that time, the roads were bad, and there was not a bell and white line on the road to support and lead blind people. Blind people had white canes made by a kind man who took branches from the mountain and shaved them down. The blind people worked at exploring with their white canes and walked from one end of the island to another in order to get treatment. They had many difficulties.

There were patients with no fingers. If they had fingers, they were dangling from paralysis or they did not have all five. Now I regret that I did not support them more at the time. Because now my fingers are short, and I am an inconvenience, just like how those patients with no fingers were inconvenient. At that time, because I could still sew with a needle, I was not inconvenienced. I could not understand their inconvenience. And at that time, when we had wounds from washing our clothes or dishes, the doctor cut down the finger immediately, like cutting a radish. A foot was changed out for a prosthetic leg immediately. It was not serious.

Temporarily coming back home from the leprosarium was real difficult at that time. We had to write many difficult things in our petition and submit it and get permission from the chief at the leprosarium. Permission was only for 1 week, so after going home, we submitted another petition and asked to extend it to 2 weeks. We used the local train and it took time, so we could only stay about 10 days, even if we had permission for 2 weeks. If we were late coming back, we had to enter a jail cell. If the returning person did not return, their guarantor had to enter the repentance room. Even if we came back late after all, the guarantor still could not leave the room. It was like a prison: There was a high concrete wall with two cages in it. The repentance room was next to the cage and had a tatami mat. I didn't want to enter the cage, so I came back by the appointed time.

In 1941, one male who was a former soldier came back late, and the staff chased and caught him, and hit his body with stones while he hung from his arms and legs. I saw it from the girls' room. I was disgusted and afraid, because they hit his head with a stone.

(Omission of middle part)

http://dx.doi.org/10.5772/65574

I think my enforced isolation was not severe, and I lived a long time by entering this leprosarium. There was a little benefit to the law. The law was abolished at1996. After abolishing the law, since all the survivors don't have *Mycobacterium leprae* anymore, we can go to the mainland easily and majestically and be accepted in the general hospital, just like everyone else. Before the law was abolished, we went shopping in a nearby town, and the townspeople would drop our change in order to avoid touching us. They were truly scared of us. Now we are given humane treatment and enjoy shopping. The old leprosarium was a scary place for people, but now it is an open-minded place, and many visitors come in. Although there are now less than two thousand survivors in 13 sanatoria, many supporters have campaigned for enlightenment, and I am very glad of it.

I think patients' families feel better about not having to worry about appearing in public with patients, because the law was abolished and we are without the offending bacteria. Although there are many survivors without family, like me.

Listener: Since this narrative will be published, what else do you want to say?

Mrs.Kadowaki: Although I am not certain, there may be new patients in the future, and I hope these future patients and their families can live ordinary live without difficulties, because in the past, the families had many difficulties. At the time, I was only a child when I got Hansen's disease, and my family was very worried about public appearances. Our surroundings were a real worry to a small patient's mother, who went crazy with worry. There was a family that had several patients in one family; how painful that must have been for the family. It was very hard for us with only one patient in the family. Although the law was abolished, we cannot return to our hometowns, but at least now we can be treated humanely. We hope sincerely that our families can live in society with peace of mind.

This next story is about how I lost my eyesight. I had a high fever due to erythema nodosum leprosum, which I got the year after I entered the leprosarium. I was very tired from caring for several severe patients as my enforced labor, and I was hospitalized several times. Because I was not getting proper treatment for *Mycobacterium leprae*, the disease was getting worse. I think the disease was getting better in the leprosarium, but it got worse in the hospital.

Before the war, we tried a new drug because the chief of the leprosarium heard good things, but it backfired, and many patients died from the side effects. I did not die, but the disease progression was getting worse. I would soon need a tracheotomy, and most importantly, I lost my hair. I was 20 years old and felt very pitiful.

As my condition was getting worse, I thought I would die if I didn't have a strong mind. Promin come to our country in 1948. At first I did not hope, because I had learned my lesson from the former side effects, but now I think I was foolish not to hope. The next year I used Promin and got better and lost the edema in my face, limbs, and larynx. I escaped having a tracheotomy, my wounds healed, and my complexion got better. These good times continued for 3 or 4 years.

Around 1953, we started a new medical treatment using drugs for *Tubercle bacillus*, which is similar to *Mychobacterium leprae*. My good condition had continued while using Promin, but I developed neuralgia, and thought, "If I cannot use my hand due to neuralgia, I will be sad," so I tried the new treatment. Now I regret this decision.

The side effects were very terrible, although I did not take large amounts of the drug, but the condition of my disease got worse again, and I lost my eyesight in both eyes for half a year. At first, my visual power decreased; next I could not open eyes because of the dazzling light and my pupils had become small like the point of a needle. I fell into a dense fog. I had pain in my eyes, but I did not feel the pain, because I was more sad for my lost eyesight. Then I lost my eyesight completely.

Slowly my eyesight got better, but then I started vomiting and I had to have enucleation of the eyeball. I was very sad. Now I cannot even wear a glass eye because of lagophathalmos, so I do not have eyeballs. If I still had my eyeballs now, I could still hope for a recovery of my visual power, but I do not have hope that medical treatment could develop new eyeballs. As a side effect, my condition rapidly got worse. My face got ugly. When my parents came to see me, they were surprised and backed away. You (listener) were surprised to see such severe sequalae when you started working here. I heard they took out my eyeballs because of glaucoma, though the doctor did not say it. The doctor did not measure the intraocular pressure, he just pushed the eyeball by hand. Although there was treatment for glaucoma at that time, the doctor's concern was only for new treatment of Hansen's disease.

I fell down to the bottom of hell when I lost my eyesight. I felt as if I were dead; I call it feeling hell on earth, with how painful it was. My friends, husband, and the association of blind people supported me for a long time. Now that my friends and husband are dead, I feel so obliged to them. When I was hospitalized, my husband cooked and encouraged me to eat. People around us said, "It is very pitiable that he has to take care of a blind wife." Losing my eyesight was sad for him, too. He took care of me though, so that I could live a long life. In 2008, he had kidney disease and hemodialysis and said, "Because I cannot eat my favorite foods, life is not as pleasant." I pitied him, but he endured for 3 years. Maybe he was preparing me to lose my husband for those 3 years. So when he died, I could endure it without crying in front of everyone, persuading myself to believe "He is still alive."

When I had good eyesight, I liked sewing, and made protectors for handicapped people to prevent wounds, and cotton work gloves with only one finger. I have lived 58 years without eyesight, and I can go to the bathroom and everywhere else by using my mouth. My hands can feel nothing, due to sensory nerve paralysis. It is as if I had burn wounds. My feet are the same. My mouth is the only part I have left that can still feel. So I wear a cloth over my mouth when I use the toilet and check for the button using my mouth.
Source: Transferred and modified from: Makiko kondo, Oshima Seisho-en: Life review of Hansen's disease survivors living in Oshima Seisho-en, 423–442, Kazama shobo, Tokyo, 2015.

Table 1. Part of the life of Ms. Hanako Kadowaki.

Hansen's disease, also known as leprosy, is an infection caused by *Mycobacterium leprae* (*M. leprae*). It has been stigmatized since ancient times due to the infected person's changed appearance caused by the deformation, and even loss, of parts of the face and limbs. After developing the first successful treatment, Promin, in 1943, it became possible to recover from Hansen's disease completely. Today, there are no new patients in developed countries. Fourteen countries, all of which are in the developing world, contain 95% of new cases, and 81% of new patients are found in India, Brazil, and Indonesia [2].

In Japan, there were many Hansen's disease patients before the end of World War II. Many of them could not live in their hometowns, so they became homeless or were forcibly placed in leprosaria with the passing of the Leprosy Prevention Law. Unfortunately, World War II was a time of national crisis in Japan, and many patients in leprosaria were treated with little regard for their dignity and human rights. In addition, many Hansen's disease survivors never had the opportunity to return to general society and had to live out their lives in sanatoriums, because the Leprosy Prevention Law, which declared that Hansen's disease patients must be isolated for their entire lives, has only recently been abolished.

We are working on continued research on nursing practices for survivors. This study explores the bioethical implications of survivors' experiences. In the second section, we discuss the lessons we can learn from their experiences. In the third section, we share the insight they have acquired by living through hardship and discuss how to genuinely restore their rights, considering that their dignity has been threatened. In the fourth section, we discuss current issues affecting survivors and how our work needs to embody the principles of bioethics through nursing practices. In the final section, we summarize what we have learned from the survivors' experiences.

2. The old leprosaria and how Japan's medical administration threatened bioethical principles

2.1. About Hansen's disease

The first characteristic [3, 4] of Hansen's disease is that the peripheral nerves are invaded, and the skin of the legs, hands, and face is deformed because the optimum temperature of the

http://dx.doi.org/10.5772/65574

Mycobacterium leprae is lower than human body temperature. The *M. leprae* cannot move into the main organs, which are located deeper in the body and thus have a higher temperature. As a result, Hansen's disease patients and survivors continue to live with a changing appearance for many years. In addition, *M. leprae* can cause people to lose their hair, eyebrows, and eyelashes; in addition, they can develop erythema nodosum leprosum, a condition that gives patients a bumpy, lion-like face.

The second feature [3, 4] consists of the multiple symptoms caused by peripheral nerve damage. These symptoms are conspicuous and cause curiosity in non-afflicted individuals. Sequelae remain throughout a survivor's life and greatly impact their quality of life, even today. Sequelae include: the inability to feel pain or notice injuries due to sensory nerve paralysis; infections from general bacteria caused by an external wound that can result in cellulitis, sequestrum, and even limb loss; motor nerve paralysis due to hand or leg deformations such as drooping hand, monkey hand, claw hand, drooping leg, and claw toe; facial nerve paralysis such as lagophthalmos, drooping eyebrows and eyelids, drooping lips, and deformation of the lips; atrophy of the nasal septum and nasal choncha, resulting in nose deformations such as saddle nose or flat nose; easily injured corneas due to lagophthalmoses, the inability to close the eye, caused by facial nerve paralysis; loss of sensitivity within the cornea due to the trigeminal nerve's inability to feel pain, which can eventually cause blindness; iridocyclitis (i.e., inflammation of the iris) can lead to blindness and eventual enucleation of the eyeball; and finally, autonomic nerve disorders can cause sweating disorders, problems with heat retention, and neuralgia in the head and face. These sequelae cause additional pain and suffering for patients.

Treatment of Hansen's disease with Promin began in the USA in 1943. In Japan, Dr. MorizoIshidate succeeded in producing Promin 1 year after World War II ended [5]. Today, the World Health Organization (WHO) promotes combination chemotherapy in terms of treating Hansen's disease [6]. Since Promin was developed, patients can achieve complete recovery without any sequelae.

2.2. The history of Hansen's disease in Japan

2.2.1. From ancient times to the early modern era

Historically, Hansen's disease [7, 8] has been one of the most frightening ailments around and was stigmatized throughout the ancient world. It appeared in the oldest civilizations of China, Egypt, and India. The first known written reference to leprosy appeared on an Egyptian papyrus document written around 1550 BC. In Japan, it was described in numerous ancient texts such as *Nihon-shoki* (AC720) and *Taiho-ritsuryo* (AC701). Based on ideas about impurity, Hansen's disease patients were regarded as impure and unclean, and the illness was seen as divine punishment for retributive justice. Furthermore, because multiple cases often occurred within the same family, Hansen's disease was thought to be genetic.

In Japanese history, patients withdrew into their homes so as not to stand out, or they coexisted with the lower classes. If they could not live in their hometowns, they often traveled on pilgrimages and slept under the eaves of shrines and temples; they had to beg because they were homeless. For example, there are 88 temples along the1200 km pilgrimage route on the island of Shikoku. Pilgrims start at the first temple until they arrive at the 88th and then return to the first one. Since the Buddhist high priest Kukai made this journey in the nineth century, pilgrims thought they were working with him at a spiritual level. People living on Shikoku thought pilgrims were an incarnation of Kukai and delightedly gave them charity. If pilgrims died by the roadside, local people buried them.

2.2.2. The beginning of the modern period

Dr. Armauer Hansen first discovered the pathogenic bacteria *M. leprae* in 1873. The old Japanese government enacted the Leprosy Prevention Law in 1909 (Act No. 11, Rai-yobo-ni-kansuru-ken) to establish five leprosaria across the nation, so that wandering and homeless patients could have a place to live. In 1931, the Leprosy Prevention Law (Act No. 58, old Rai-yobo-ho) was amended, and local governments started campaign to find all patients and forcibly send those living in their homes to leprosaria. This campaign caused local inhabitants to think of Hansen's disease as a terrifying epidemic; as a result, patients were ostracized in their hometowns and villages [9, 10].

With World War II coming, Japan faced a national crisis as modernization picked up and wealth and military strength increased. With the outbreak of war, Hansen's disease was regarded as a national disgrace. This is because (1) homeless, begging, and wandering patients were a symbol of Japan not being a civilized country; (2) based on ideas of ethnic cleansing, weak individuals (such as those with low intelligence or schizophrenia, those who were physically disabled, and those who had Hansen's disease or tuberculosis) were ostracized; and (3) since Hansen's disease occurred frequently in young males, these patients could not become soldiers and contribute to the nation's military force. As a result, Hansen's disease was viewed as a terrifying epidemic and national disgrace, and the government proceeded to enforce policies of lifelong confinement and isolation [11, 12].

The Leprosy Prevention Law was renewed in 1953 (Act No. 24, new Rai-yobo-ho), even though the National Leprosaria Residents' Council (ZenkokuKokuritsu Rai Ryoyo-shoKan-jyaKyogikai) opposed it on the grounds that lifelong isolation threatened patients' human rights [13, 14]. By this point, World War II was over, reconstruction under a new democratic constitution was under way, Promin had been developed, the illness could now be cured, and the WHO recommended that the government reconsider its segregation policy.

The Leprosy Prevention Law was renewed in 1953 with the addendum, "We must reconsider this policy as soon as possible" [15]. However, the law remained in place until 1996, although legal regulations were alleviated through patients' efforts. Patients created self-government associations at each leprosarium, which cooperated with all other leprosaria and negotiated with the Ministry of Health, Labor, and Welfare. In the 1950s, they obtained much-needed medical treatment and pensions for all patients. They ended requirements for patient labor and began employing leprosarium staff in the 1960s, and improved housing for patients in the 1970s. Patient living conditions got better gradually, but these improvements had long been needed [13, 14, 16].

http://dx.doi.org/10.5772/65574

In 1996, the Leprosy Prevention Law was finally abolished. Survivors won a lawsuit striking down national indemnity, and the court convicted the Japanese government of enacting an unconstitutional law. The nation officially apologized, with formal apologies from the Prime Minister, the National Diet, the Minister of Health, Labor, and Welfare, and the Chief Justice of the Supreme Court. In 2009, the government established a new law, the Act to Accelerate the Resolution of Problems of Hansen's Disease (Act No. 82, Hansenbyomondai no kaiketsu no sokusin ni kansuru horitsu). The new legislation guaranteed that survivors could live at the remaining sanatoria for the rest of their lives, would be given a comfortable life and medical care, would be aided in their social lives and rehabilitated into mainstream society, receive help to restore their reputations, be buried with dignity, and that their relatives would also be supported [17].

2.3. The circumstances of leprosaria under the segregation policy

Although the treatment of Hansen's disease patients has slowly improved, it was inhumane from the time that the first Japanese leprosarium opened in 1907 until the end of World War II [13, 14, 18]. For example:

- In the early days, the director of a leprosarium was not only a doctor, but also a policeman. The police were in charge of enforcing confinement.
- The Leprosy Prevention Law did not allow patients to leave the leprosaria, and they had to remain there their entire lives. Although the leprosaria were medical institutions, they also had crematoria, graveyards, and religious buildings (**Figure 1**).
- In order to prevent patients from escaping, their property was converted into a currency that could only be used within the leprosaria.
- When patients entered a leprosarium, they were compelled to take on a new name and could not use their real one.
- The chief of a leprosarium had a great deal of power, and if patients did not obey the rules or resisted authority, the chief could punish them or send them to solitary confinement at his own discretion.
- Patients were permitted to marry, but not to have children. Before any marriage could take place between patients, the male was required to have a vasectomy.
- When patients entered a leprosarium, they had to sign a letter that stated they accepted to be dissected after death. **Figure 2** shows an artistic depiction of a dissection table used in certain leprosaria. This table was thrown into the sea, but washed back up on shore. Now, the table is on display on the island as a symbol of human rights violations.
- In order to operate leprosaria with as few doctors and nurses as possible, patients were required to work. In 1949, there were 62 doctors and 253 nurses in all leprosaria in the country, while there were 8318 patients. Patients worked various jobs such as caring for severely disabled patients, as medical assistants, conducting funerals for deceased patients, washing, working on engineering projects for the leprosaria, or carrying items for staff or disabled patients.

- The space for patients and the area for those without Hansen's disease were clearly divided. Patients could not enter the disease-free zone, and if a doctor needed to enter a patient's home due to severe illness or injury, the doctor wore rubber boots, despite the Japanese custom of not wearing shoes inside the home.

Figure 1. There are graves of patients in spite of medical institution (Picture provided by Oshima Seisho-en©).

Figure 2. The dissection table used in certain leprosaria, which was thrown into the sea, but washed back up on shore (Illustration by Akika KONDO©).

2.4. Introduction to the study: why did Hansen's disease patients lose their limbs?

In this section, we introduce our study [19], discuss the factors that caused patients to lose their limbs, and examine how patients' daily lives were affected by this occurrence.

<Categories>	Subcategories
<1. Loss of pain as a caution>	(1) I moved too much, because I did not feel pain from my wounds
	(2) I could not notice a heavy burn, even if my skin was charred, because I felt numb
<2. Carrying loads in the same region>	(3) I protected a region of my body that was paralyzed and deformed, so I carried my loads at same region and caused new wounds
<3. Collapsing from within the body>	(4) The bone went stale, and my finger disappeared
	(5) A tumor mass from leprosy erthema nodosum was crushed naturally
<4. Labor is vital because of poverty>	(6) Because I was poor, I could not live without working
	(7) I could not miss work due to perennial wounds
<5. Poor living conditions promoting worsening of wounds>	(8) Because we did not have transportation, I had to walk home from the treatment room, and the bandage came off immediately after treatment
	(9) I was wounded easily, because I worked hard wearing wooden clogs on an unpaved road
<6. Contaminated wounds and lack of hygiene>	(10) Because I could not avoid scrubbing, washing, and field work, my wound was contaminated immediately
	(11) Because gauze, bandages, and drugs were too expensive, I did not have the means to disinfect my wounds when they were dirty from muddy work
	(12) Tweezers and gauze were hard to get
	(13) There were many flies in our food, and maggots breed in the wound
<7. A poor medical system>	(14) Because there were many patients, I was too busy and could not wait to be seen by the doctor
	(15) The doctor did not do house calls, even if we had a serious condition with a high fever or tumor mass from leprosy erythema nodosum
	(16) Because there was no doctor, my wound was treated too late and got worse
	(17) The doctor did not treat me like a human
	(18) I had a bitter experience being experimented on, so I became timid of medicine
<8. Inadequate treatment by untrained individuals>	(19) Patients did amateur wound treatment, so they weren't treated correctly
	(20) Because I did my wound treatment myself, it became infected
<9. Superstition>	(21) Because I held the superstition, "If we cure perennial wounds, we will lose our eyesight," I deliberately created new wounds

Source: Transferred and modified from Ref. [19].

Table 2. Causes of repeated wounds.

This study uses aging survivors' narratives to clarify why Hansen's disease patients had repeated wounds, which led them to lose their limbs, and how they coped with these afflictions until starting Promin treatment. After starting to take Promin, patients no longer developed multiple severe disorders that resulted in new wounds or lost limbs, although they still had preexisting severe sequelae that continued to negatively influence their quality of life.

[Categories]	Subcategories
[A. Acquiring their own style of treatment for perennial wounds]	(a) I checked temperature of people or things with my tongue, which was the only part of my body that could feel hot or cold, as everything else had sensory nerve paralysis
	(b) Because my wound was small at surface, but was deep, I did not know how bad it was by looking at it, so I made a judgment whether I needed treatment or could use a bubble of hydrogen peroxide
	(c) Before my wounds got worse, I started using my own style of wound treatment
	(d) I scraped off necrotized parts of skin without mercy
	(e) I burned tweezers in the fire of a candle and dug into the wound
	(f) I stitched a crack in my skin at finger joint closed by sewing it with yarn myself
	(g) I blended drugs to make effective drugs to either protect or remove skin
	(h) I made drugs from horse teeth
	(i) Because I did not have drugs, I beguiled my neuralgia by continuing to walk around during the middle of the night
	(j) I protected the places that wounded easily
	(k) I devised a way of walking in order to avoid creating a new wound
	(l) I devised a way of not getting my wounds wet or dirty
	(m) I stretched and fixed crooked fingers, in order to prevent blood blisters
[B. Not being able to cure wounds by slack treatment]	(n) I knew I could not be cured by slack treatment
	(o) I was particular about my own style of treatment
[C. Selecting a therapist based on the severity of the wound]	(p) I went to the doctor only when my wound was beyond the control of the patients
	(q) I relied on the nurses, who were getting used to wound treatment, had good skills, and were improving their skills by evaluating the patients
	(r) Patients managed all medical treatments
[D. Relieving suffering through limb amputation]	(s) I allowed my wound treatment to cause atrophy
	(t) I selected cutting off a hand, foot, or finger in order to relieve suffering in the wound
[E. Giving priority to labor, and even choosing to cut off limbs]	(u) I selected cutting off a limb in order to continue working and sustain the lives of my wife and children who live outside of the leprosarium
	(v) I had no choice but to work in order to sustain my own life, so I selected cutting off my own limb
[F. Keeping a limb due to not being able to work or having a work release]	(w) My limbs remain because I could not work due to lost eyesight or neuralgia
	(x) I was supplied from home, so I did not need to work. Thus, my limbs remained

Source: Transferred and modified from Ref. [19].

Table 3. Coping with repeated wounds.

Since Promin was developed during World War II and didn't arrive in Japan until after the war ended, severe sequelae occurred when the entire nation was suffering from the effects of poverty. Our results show that limb loss during this period was caused not only by pathological characteristics such as sensory and motor nerve paralysis in the limbs, but also by a shortage of proper medical care and hygiene, in addition to economic distress. The findings show how patients lived with Hansen's disease and what multiple severe disorders, wounds, and limb loss meant to patients living in poverty.

Narrative data were obtained from ten survivors and analyzed using a qualitative and inductive method. The average age of the survivors interviewed was 81.8 ± 2.7. Nine individuals had the lepromatous disease type, while one person had the tuberculoid type. Eight individuals had limb deficiencies, nine had hand and leg drooping, three were blind, and all ten had both contracture and neuralgia.

There were nine main causes of repeated wounds (see **Table 2**) and six main ways of coping with them (see **Table 3**). **Figure 3** shows the relationships between the cause categories and coping categories, and reveals that there were three ***core categories*** of relationships. The first core category is the ***negative spiral inducing a deterioration of wounds***, as shown in **Figure 3** with a blue line. The second core category is ***living daily life with perennial wounds in poverty***, shown with a red line. The third category ***is severe sequelae as compensation to survive poverty***, shown with a green line. In **Figure 3**, <> indicates the cause categories, [ ] indicates the coping method categories, indicates supplementary explanations for wounds, and × indicates a disruption in the process.

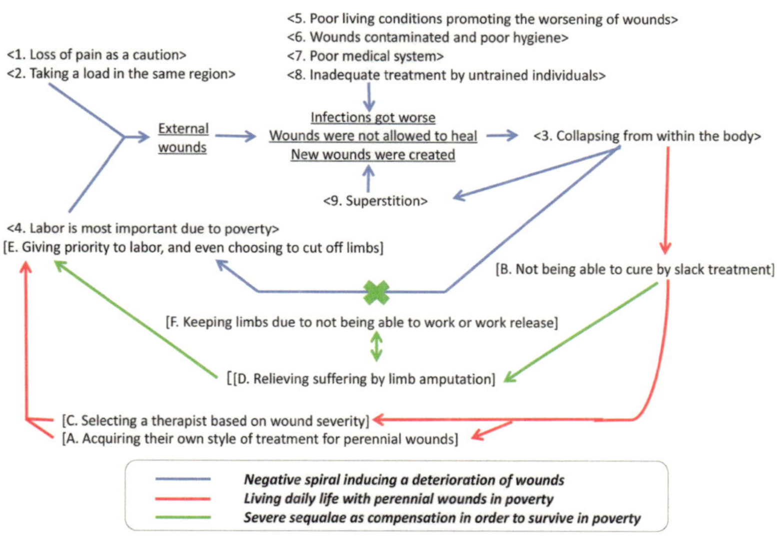

Figure 3. The cycle of repeated wounds among Hansen's disease survivors. Transferred from: Yamaberi et al. [19].

2.4.1. *The negative spiral that caused wounds to deteriorate*

Hansen's disease patients were wounded easily due to <1. loss of pain; pain signals that a person is hurt> and repeatedly <2. carrying loads in the same region > to compensate for other deformities or paralysis. However, patients also believed that <4. labor is vital because of poverty>. They continued to work under these circumstances, leading to increased external lesions. Furthermore, due to <5. poor living conditions that worsen wounds>, <6. contaminated wounds and a lack of hygiene>, a <7. poor medical system>, and <8. inadequate treatment by untrained individuals>, existing infections got worse or became infected, wounded areas were not treated or allowed to heal properly, new wounds formed, and limbs deteriorated until they finally <3. collapsed from within the body>. However, many patients continued to [E. give priority to labor, even choosing to cut off their own limbs], and kept working in spite of their wounds, due to <1. loss of pain>. Thus, their wounds became worse, and they fell into a negative spiral. On the other hand, some patients were afraid to lose their eyesight and believed in <9. superstition>, such as: "If you recover from your wounds, you will lose your eyesight." These patients created new wounds on their bodies to avoid fully healing and thus not lose their eyesight. No matter the causes of repeated wounds, it led patients to fall into a negative spiral of deteriorating lesions.

2.4.2. *Living with perennial wounds in poverty*

Since patients fell into these negative spirals, having lesions was a matter of course, so they called them "perennial wounds." They knew empirically that they [B. could not cure their wounds with slack treatment], so they [A. acquired their own style of treatment for perennial wounds] and routinely attempted to heal themselves. If patients required specific medical treatment, they [C. selected a therapist based on the severity of the wound]. For example, when more severe wounds required cutting off a large bone, patients sought a doctor's help. If a less severe wound required cutting off a small bone at the finger or toe, they often cut it off by themselves or went to a nurse. Thus, patients made decisions about and arrangements for treatment themselves. They received support from nurses, except in cases of intractable severe gashes, when they sought the aid of doctors. With these coping techniques, patients barely maintained their lives in severe poverty, continuing to [E. give priority to labor, even choosing to cut off their own limbs]. Thus, they lived with "perennial wounds."

2.4.3. *Severe sequelae as compensation to survive poverty*

Although patients could cope with ordinary external wounds by [A. acquiring their own style of treatment for perennial wounds], the wounds <3. collapsing from within the body> were intractable, and they [B. could not cure their wounds with slack treatment]. Patients often chose to amputate limbs to return to work more quickly and easily; they had to work hard to support themselves or send money back to their wives and children who lived outside the leprosarium. Others hoped amputation would [D. relieve their suffering]. On the other hand, when patients could not work (e.g., due to loss of eyesight), the negative spiral was broken. As a result, patients [F. kept their limbs due to not being able to work or having a work release]. Both the loss of eyesight and cutoff limbs meant that patients would live with severe sequelae as compensation to survive poverty.

2.5. What lessons can we learn from the old leprosaria?

Above, we introduced historical facts about the Hansen's disease isolation policy and showed why and how patients contracted multiple severe sequelae. In this section, we will discuss what lessons we learned (see **Table 4**).

http://dx.doi.org/10.5772/65574

Actions		Threats to the four principles of bioethics		Respect/unavoidable principle of bioethics	
		Reasons the bioethical principles are threatened	**Threatened principle(s)**	**Results of respect or unavoidable principle of bioethics**	**Respected/ unavoidable principle**
(1) Enforced confinement and lifelong isolation					
Enforced confinement	• Protecting homeless patients in leprosarium	—	—	• Protect people in crises (such as the homeless)	beneficence
	• Finding patients in their homes, and enforcing confinement using police power	Exercising compelling power (depriving freedom of decision-making)	Respect for autonomy	—	—
	• Building leprosaria in hard-to-reach places in order to prevent patients' escape	Exercising compelling power (depriving freedom of decision-making)	Respect for autonomy	—	—
Lifelong isolation	• Separating patients from their families	Plundering good (living with their family) from the patient's life	Non-maleficence	—	—
	• Plundering their right to live in society	Plundering the opportunity to contribute to society or the family	Non-maleficence	—	—
	• Causing emotional suffering and thoughts of suicide	Creating existential suffering	Non-maleficence	—	—
	• Actual harm done to families: by patients' forced isolation, the family lost to live with their husband/wife, son/daughter, father/mother, brother/sister, etc. If patient was sustained family's life by his income, the family had difficulties with economy.	Plundering good (living with family, income) from a family	Non-maleficence	—	—
Isolation based on weak medical evidence	• Enforced confinement for a disease with weak infection power and little need for isolation	Impelling unnecessary isolation (physical restraint)	Non-maleficence	—	—
	• Prioritizing the nation's convenience over patients' needs without sound medical evidence	Neglecting the principle of medical care which determines the necessity of hospitalization based on patients' medical condition	Beneficence	—	—

Patients shouldering this burden unilaterally	• The burden of isolation was solely on patients in order to prioritize the nation's needs to prevent infection, empower the wartime regime, and enhance the national reputation	Imposed burden only on patients due to national policy	Justice and/or equality	—	—
	• Patients endured lifelong isolation to protect the general public from infection	Imposed burden only on patients in order to control infection.	Justice and/or equality	—	—
Failing to abolish the Leprosy Prevention Law	• By fighting patients' association against government, they were permitted temporary going out, but they must prove to disappear mycobacterium leprae into their body and got permission to going out from head of leprosaria.	—	—	• Defend individuals' freedom of action through flexible application of the law	beneficence
	• The government took many years to abolish an unnecessary law that violated human rights	No one in law, medicine, or administration took action	Beneficence	—	—
	• Due to this law, survivors continue to suffer from the old isolation policy	They continued to suffer unnecessarily	Non-maleficence	—	—
	• Patients lost years of opportunity to come back into society, and survivors were elderly when the law was finally abolished	Plundering the good of living in society	Non-maleficence	—	—
Forming negative public opinion	• The leprosaria protected patients from discrimination and persecution	—	Non-maleficence	• Use national power to protect the weak from harm, persecution, and discrimination	beneficence
	• Promoted discrimination and persecution against Hansen's disease patients among the general population through the campaigns to find patients in their homes and forcibly send them to leprosaria	Promote suffering and discrimination) indirectly	Non-maleficence	—	—
	• Spread discrimination and persecution from patients to their families, due to the campaigns to	Spread suffering and discrimination for family and relatives of patients	Non-maleficence	—	—

http://dx.doi.org/10.5772/65574

	find patients in their homes and forcibly send them to leprosaria				
Avoid spreading information about the living conditions of patients to the public	• The general public did not know about the poor conditions within the leprosaria	The public did not have the opportunity to get important information and make decisions based on good, common sense	Beneficence	—	—
(2) Poor treatment of patients in leprosaria (a) Direct threats					
Vasectomy and abortion	• Abortion	Plundering the life of a child	Non-maleficence	—	—
	• Child bearing was not allowed	Plundering good from a patient's life, such as becoming a parent, the delight of child bearing, and enjoying old age surrounded by child and offspring	Non-maleficence	—	—
	• Patients were not allowed to refuse this treatment	Exercising compelling power and depriving freedom of choice	Respect for autonomy	—	—
	• Sorrow for the loss of a child, sense of humiliation, and shame for the treatment	Psychological laceration	Non-maleficence	—	—
	• The invasive operation itself	Physical laceration	Non-maleficence	—	—
Pathological dissection	• When entering the leprosarium, all patients were required to sign a consent form permitting dissection after death	Exercising compelling power and depriving freedom of choice	Respect for autonomy	—	—
	• Desecration of the dead	Harm the dignity of the dead	Non-maleficence	—	—
	• Medical contribution	—	—	• Elucidate pathological cause of disease	beneficence
Chief at leprosarium held strong power	• Patients had to comply with the chief's orders	Exercising compelling power and depriving freedom of choice	Respect for autonomy	—	—
	• Punishment resulting in physical pain	Causing physical pain	Non-maleficence	—	—
	• Punishment resulting in weakness or death	Plundering a patient's life	Non-maleficence	—	—
	• This power was a warning to other patients, in order that they submit to orders meekly	Exercising compelling power and plundering self-motivation from the majority of patients	Respect for autonomy	—	—

Enforced exchange of property to money that could only be used in the leprosarium	• Forfeiture of property	Plunder economic power	Non-maleficence	—	—
	• Preventing escape by depriving patients of cash	Exercising compelling power and depriving freedom of choice	Respect for autonomy	—	—
Enforced adoption of a new name and giving up one's real name	• Forfeiture of one's own name	Forfeiture of the social life patients had prior to entering the leprosarium and erasing the patient from society	Non-maleficence	—	—
	• Denying patients their identity	Denying personal uniqueness and individuality	Respect for autonomy	—	—
Dividing the leprosaria between patient areas and healthy person areas	• Discriminatory treatment between patients and staff	Cause suffering through discriminatory treatment	Non-maleficence	—	—
	• The purpose of the division was to prevent infection of the medical staff	—	—	• Infection control	beneficence not for the patient, but for medical staff
Enforced labor	• Patients did not have the right of refuse	Exercising compelling power and depriving freedom of choice	Respect for autonomy	—	—
	• Heavy physical labor	Causing physical pain	Non-maleficence	—	—
	• Poor wages for hard work	Unfair exploitation	Non-maleficence	—	—
	• The leprosarium was not able to sustain itself without patients' labor	Only patients had to shoulder the burden of sustaining the leprosarium	Justice and/or equality	—	—
	• Patients were not given suitable care but were used as a labor force	Received poor medical treatment	Beneficence	—	—
(b) Indirect threats					
Conditions caused multiple severe sequalae	• The main cause of limb loss was the pathological characteristics of Hansen's disease	—	—	• Limb loss was unavoidable due to the pathological characteristics of the disease	non-maleficence
	• Severe poverty meant the patients could not live without working, even with wounds	Intentional, willful negligence	Non-maleficence	—	—
	• Even with lost limbs, patients had to work. The only exceptions	Patients could not chose to rest their	Respect for autonomy	—	—

http://dx.doi.org/10.5772/65574

	were for the blind or individuals with another severe medical condition	own bodies, except when they had a severe medical condition			
	• Poor infrastructure and an unclean environment in caused wounds to worsen and patients to lose limbs	Poor physical environment did not protect from mixed infection and additional wounds	Beneficence	—	—
	• There was shortage of medical staff and goods	Received poor medical treatment	Beneficence	—	—
	• New drugs were before there was evidence of their therapeutic effect, so more sequalae occurred	Non-therapeutic drugs were tested on the human body	Non-maleficence	A cure was eventually developed with new drugs	beneficence
	• The entire nation struggled with poverty, war, and earthquakes	—	—	• The bad environment was unavoidable, considering social situation at that time	justice and/or equality
	• Multiple severe sequalae caused by enforced work continues to decrease patients QOL even no.	Causing lifelong, irreversible harm	Non-maleficence	—	—
Creating negative public opinion	• Patients' changing appearance was a source of discrimination and was encouraged by the government's policy	Promoting harm through discrimination	Non-maleficence	—	—

Table 4. The mistakes of isolation policy for Hansen's disease patients through the four bioethical principles.

2.5.1. *The four bioethical principles*

The foundation of biomedical ethics goes back to the Hippocratic Oath in BC 54. After World War II, the Nuremberg Code was created in 1947, following the consequences of the war and the Holocaust caused by the Nazis; this set of research ethics was developed to protect subjects' will and freedom in future medical studies [20]. In 1964, the World Medical Association (WMA) announced the Declaration of Helsinki, which outlines rules of medical research as based on the Nuremberg Code; it continued to revision [21]. In 1979, the Belmont Report was created by the National Commission for the Protection of Human Subjects of Biomedical and Behavioral Research and illustrates principles of medical ethics to protect research subjects [22]. The WMA announced the Declaration of Lisbon in 1981, which describes patients' rights [23]. The Council for International Organizations of Medical Sciences (CIOMS) created the International

Guidelines of Biomedical Research Ethics in 2002 [24]. The Belmont Report clarifies biomedical ethics principles such as respect for people, informed consent, beneficence (i.e., evaluating benefits and risks), and justice (selecting subjects). The creation of biomedical ethics placed limits on human experiments and spread throughout the research world and medical practice.

Japan has a long history of medical knowledge, skills, and ethics. The medical book *Ishinbo* was written by Dr. YasuyoriTanba for Emperor Enyuin AD 984 [25]. After World War II, bioethical principles were imported from North America and Europe. Biomedical values from North America are based on the Belmont Report or the ideas of Beauchamp and Childress [26], who emphasize respect for autonomy, non-maleficence, beneficence, and justice. The tenets from Europe are based on the Barcelona Declaration, which was adopted at the Euro-Mediterranean Conference and underscores autonomy, dignity, integrity, and vulnerability. In terms of actual practices, modern bioethics is based on various methods such as the procedure- and narrative-based approaches [27]. In our analysis, we selected principles based on the American tradition, on which medical researchers and practitioners in Japan reached a consensus [28, 29].

The four bioethical principles are fundamental rules meant to address and resolve ethical issues in medicine. The first tenet is respect for autonomy, meaning that we must respect a patient's autonomous decision-making and to remember that informed consent is important in every situation. This belief requires that medical professionals: (1) tell the truth, (2) protect others' privacy, (3) adhere to the duty of confidentiality, (4) obtain informed consent for any kind of physical invasion, and (5) if asked, help the patient make important decisions. The second principle is beneficence, meaning that we must act for the benefit of others. This notion requires that we: (1) advocate for the rights of others, (2) protect others from harm, (3) remove risks that could cause harm, (4) support disabled persons, and (5) help individuals who are facing a crisis. The third rule of non-maleficence prohibits doing harm to others. This principle covers the following rules: (1) do not kill, (2) do not cause pain or suffering, (3) do not hinder another's ability, (4) do not create discomfort for others, and (5) do not steal or destroy the good in another's life. The fourth tenet of justice and/or equality means that social benefits and burdens must be distributed equally. This value requires that medical professionals distribute limited medical resources (such as manpower, goods, or capital) equally. We must avoid baseless discrimination and adequately balance competing requirements.

2.5.2. The mistakes of the isolation policy of Hansen's disease patients through the four bioethical principles

2.5.2.1. Enforced confinement and lifelong isolation

Enforced confinement threatened respect for personal autonomy because the government used its power via the police to compel patients to enter leprosaria and prevent them from escaping. On the other hand, homeless, wandering patients who entered leprosaria experienced a more comfortable life and improved medical care than they had previously; in this way, the leprosaria contributed to beneficence for some patients.

Lifelong isolation threatened non-maleficence because patients were denied the happiness of living with their families and exerting their own abilities in society. Patients' families were robbed of important family members and of patients' contributions to the families' incomes; family members could not live with patients throughout their lives, and if a male patient formerly supported his wife and child, his family lost income after he entered a leprosarium.

In addition, because the required isolation was based on weak medical evidence, it also threatened non-maleficence in the sense that patients had to endure unnecessary seclusion, even though Hansen's disease was not highly infectious. It also endangered beneficence because the country's circumstances—namely the idea that Hansen's disease patients were a national disgrace and not acceptable potential soldiers for the war effort—preceded medical requirements. Patients shouldered the burden of lifelong isolation in order to fulfill the nation's policy of controlling infection, and the rights of the state were considered more important than those of patients; thus, justice and/or equality was also breached.

Delaying in nullification of the Leprosy Prevention Law violated beneficence because no one in the fields of law, medicine, or local and national government took action to change this regulation for a long time. However, operational relaxation of the law did allow patients to leave the leprosaria occasionally and contributed to beneficence a little. By delaying abolishment of the law, non-maleficence was violated because patients continued to suffer unnecessary isolation. They also lost years of opportunity to come back to general society. When the law was finally annulled, the remaining patients were almost all elderly.

Enforced confinement by police and the campaign to send patients to leprosaria influenced public opinion and promoted fear of the illness. It led the general public to expel patients from communities and caused families to discriminate against their sick relatives, even though entering leprosaria was supposed to protect patients from public prejudice and persecution. All of this violated non-maleficence.

The lack of information about the realities of life within leprosaria also jeopardized beneficence. Except for some supporters, the general public did not know about or concern themselves with leprosaria, even though many people came into leprosaria and interacted with survivors after supporting the Leprosy Prevention Law. After World War II, Japan rapidly developed both economically and democratically. If more citizens had known about the conditions within leprosaria and the history of Hansen's disease, there may have been many more advocates for patients.

2.5.2.2. Violations of patients' rights in leprosaria

Often, treating patients in leprosaria threatened their rights either directly or indirectly.

2.5.2.2.1. Direct threats

Vasectomies were enforced for men prior to getting married, and pregnant wives were forced to abort their children, which directly violated non-maleficence because these actions either prevented a fetus from becoming a child (which goes against the principle "do not kill") or took away patients' delight in becoming parents and the joy in being surrounded by children in old age. These actions caused patients to suffer the loss of being able to have children and often a sense of shame and humiliation in having such invasive operations. These surgeries also endangered respect for autonomy because patients could not refuse them.

When patients entered a leprosarium, they were required to sign a letter of acceptance regarding their dissection after death. They could not refuse. This violated respect for autonomy and non-maleficence. Even if patients refused to have their bodies dissected after passing away, their bodies were cut up anyway and their organs removed. They were not able to benefit from operations (e.g., to remove cancerous tumors). In Japan, funerals are very important rituals, and dissection without informed consent constitutes blasphemy. However, dissections did contribute to beneficence because they helped reveal the pathological cause of Hansen's disease.

The chief of a leprosarium had a great deal of power, including the ability to punish patients. This jeopardized non-maleficence because patients who did not obey the rules were imprisoned. Furthermore, this threatened respect for autonomy, because this kind of power not only violated the rights of individual patients, but also served to warn others against breaking the rules.

Changing patients' property to a currency that was only used within the leprosaria threatened both respect for autonomy and non-maleficence. Regarding the violation of respect for autonomy, the goal of this action was to prevent patients from escaping by making sure they could not use their money in the real world if they did manage to flee. In terms of threatening non-maleficence, since property was essential to one's social life, depriving patients of property made it impossible to have a normal social status.

Forcing patients to assume a new name and not allowing them to use their real one threatened non-maleficence because it meant depriving patients of a social life and "erasing" them from society. It also violated respect for autonomy because prohibiting patients from using their real names deprived them of their identities and harmed their sense of individuality.

Dividing leprosaria into areas for patients and healthy people threatened non-maleficence because it resulted in discriminatory treatment. While the purpose was to prevent medical staff and other healthy people from becoming infected, it only contributed to beneficence to help medical staff, but not patients.

Enforced, compulsory labor violated respect for autonomy and threatened non-maleficence because patients suffered physically during their labor and were not sufficiently rewarded for it. It threatened justice and/or equality because only the patients shouldered the burden of maintaining the leprosaria, and it endangered beneficence because the leprosaria did not provide sufficient medical services, despite being medical institutions.

2.5.2.2.2. Indirect threats

Indirect threats to Hansen's disease patients promoted multiple severe sequelae. In our study (see above), the main causes of repeated wounds were the loss of pain and severe poverty, while the main sub-causes were poor living conditions, a lack of proper hygiene, and a deficient medical system. Prior to World War II, there was no silver bullet to cure Hansen's disease, and the characteristics of the illness caused patients to easily develop repeated lesions. Thus, we cannot conclude that multiple severe sequelae threatened non-maleficence. However, doctors could clearly foresee that if patients were forced to work hard, they would have repeated wounds and would be forced to cut off limbs. In spite of this, the government still compelled patients to live and work in extreme poverty and did not give them the chance to rest and recover from their injuries. This situation threatened both non-maleficence and respect for autonomy. In addition, by not providing a physical environment or suitable medical care to prevent mixed infections and protect wounds, beneficence was violated. Trying new drugs on patients did contribute to beneficence by offering hope for a cure, but without evidence of any therapeutic effects or informed consent, patients continued to experience severe sequelae, thus endangering non-maleficence. Patients still suffer from multiple, irreversible sequelae today, which negatively affects survivors' quality of life; thus, non-maleficence continues to be violated today.

By allowing policies that resulted in severe sequelae, the leprosaria also indirectly violated non-maleficence, because limbs <3. collapsing from within the body> caused by enforced work in poor conditions led people to discriminate against patients and fear their changing bodies.

2.6. What lessons have we learned from old leprosaria?

There were undoubtedly some humanistic doctors, nurses, and religious men and women in the old leprosaria who cured patients and cared for them with devotion. However, the government's policy on Hansen's disease patients was wrong. We have learned the following lessons from history:

- We must create systems to ensure that medical policies do not violate bioethical principles.
- We must ensure that infection-prevention methods do not increase people's fear, especially in cases of lethal and tragic diseases. For example, Ebola hemorrhagic fever could cause people to fear and ostracize patients.
- We must accurately communicate the pathological causes of symptoms to the public; individuals are often stigmatized for symptoms that are considered unusual or mysterious. For example, limb loss by sequestrum can occur not only with Hansen's disease, but also with diabetes and thromboangiitis obliterans.
- The majority, not the persecuted minority, makes decisions about policy. If the majority is not concerned about how the minority is treated, discrimination and bullying can continue. Thus, we must ensure that the majority has correct and accurate information and that we foster a sense of justice so that persecution is not allowed to persist.

3. The wisdom of aging Hansen's disease survivors, who survived extreme situations

In the last section, we discussed the various violations of the four bioethical principles in medical policy regarding Hansen's disease. In this section, we discuss how patients survived extreme situations, and what wisdom aging survivors can pass on to future generations. We explain how they survived in detail because we believe that in order to restore human rights, we need to understand survivors' strength and resilience. Their wisdom can be seen as universal, for the benefit of all (see **Table 5**).

Categories	Subcategories
1. Sustaining everyday life through self-sufficiency	(1) There were water shortages and poor water quality, a shortage of food, poor living conditions, and information disruption
	(2) Patients called on all their wisdom and ingenuity in order to obtain clean drinking water, cash, and food
	(3) Patients endured heavy work
	(4) Patients reclaimed new land and water resources
2. The courage to survive extreme hardships	(5) Patients' difficult situations did not weigh heavily on their minds, because they found joy and fun even in hard work
	(6) Patients without heavy sequalae enjoyed having youthfulness and energy
	(7) Patients could appreciate that their difficulties were trivial, because some had previously experienced homelessness and the wandering life of a pilgrim
	(8) Patients were not bothered by their severe poverty, because all residents were equally poor
3. Resourceful people coming together	(9) Patients had held a variety of occupation before entering the leprosarium. Patients were grouped by profession, specifically by professions that were needed to sustain the leprosarium
	(10) Professional patients taught their jobs to novice patients
	(11) If there was no professional in their midst, all jobs were done by trial and error, with patients learning by imitation
	(12) Some patients who had experienced homelessness and the wandering pilgrimage had excellent wisdom for everyone coping with severe poverty in the leprosarium
	(13) Doctors treated patients equally based only on their medical judgment
4. Assembling self-sufficient organizations for self-defense	(14) There was physical and psychological distance from the leprosarium staff, who treated patients inhumanly. The government and leprosarium did little to help the patients
	(15) Patients created self-sufficient organizations, such as a young persons' association, a vigilante corps, and a women's association. The self-governing associations decided terms of payment based on the difficulty of a job, and they systematically prepared to receive pilgrim patients from outside
	(16) Since all residents were assigned jobs, even children, the associations made sure that everyone was doing the appropriate job for his or her skills and abilities
	(17) The patients' organizations created systems to stockpile food and water

Categories	Subcategories
	(18) They made rules about water use during times of scarcity
	(19) If anyone did not follow the rules, a popular leader made sure they did
	(20) When male doctors went away to fight in World War II and medical care was shorthanded, the patients undertook their own care
	(21) Strolling players helped with shopping outside of the leprosarium and gifts from families living outside the leprosarium helped alleviate patients' poverty
5. Patients created a mutualaid system	(22) Patients gathered a portion of all payments for enforced work and distributed it equally to patients who could not work due to severe sequalea such as lost eyesight or other critical conditions
	(23) They shared food and goods with everyone. If someone was in poor physical condition and could not work, somebody took their place
	(24) They were considerate of weakest among them, such as girls who were ashamed of receiving charity or begging, or blind patients who hesitated to receive care from other patients. If an individual had food, they did not eat in front of patients without
	(25) When blind patients received charity from working patients, they often gave the workers a massage as a sign of their gratitude
6. They agitated to compel the government to provide better support	(26) Some survivors decided to fight against the government and encourage the abolition of the Leprosy Prevention Law, deciding to once again be homeless and live in poverty
	(27) After receiving better support from the government, the survivors' daily lives improved dramatically

Source: Adapted from Ref. [30].

Table 5. Wisdom of aging Hansen's disease survivors, who survived in extreme situations.

3.1. Introduction to our study: how did Hansen's disease patients survive severe poverty?

This study [30] used narrative data from aging survivors to clarify how patients survived in poverty before World War II. The subjects are living in the same sanatorium (mentioned in the above study), but that study specifically selected survivors who had repeated wounds and thus contained many people with the LL (Lepromatous) type. For this research, we focused on older survivors. The average age of our subjects was 84.3 years old, and the average age for all survivors living in this sanatorium was 80.2 year sold. The subjects had lived at the sanatorium for an average of 66.3 years, and all survivors in this particular sanatorium had lived there for an average of 52.6 years. We gathered data through semi-structured interviews and analyzed it using qualitative and inductive methodologies.

The narrative data were integrated into six categories of coping methods and types of resourcefulness. < 1. Maintaining everyday life through self-sufficiency> showed that there were water shortages and poor water quality, a dearth of food, poor living conditions, and information disruption. They endured heavy labor, and reclaimed new cultivated land and water resources. They were self-sufficient and found ways of coping with a difficult daily life.

<2. The courage to survive extreme hardship> showed that patients' difficult situations did not weigh heavily on their minds because they found joy and even fun in hard work. Those without grave sequelae had youthfulness and energy. Everyone could appreciate that their hardships were trivial compared to worse situations, because some patients had previously experienced homelessness and the wandering life of a pilgrim. Their severe poverty did not bother them because all residents were equally poor. This showed that these patients had the psychological strength to overcome hard labor, illness, poverty, and still have a positive outlook.

The wisdom of <3. resourceful people coming together> was important because patients had held a variety of occupations prior to entering the leprosarium. Patients were grouped by profession, specifically by vocations that were needed to sustain the leprosarium. Professional patients taught their jobs to novice ones, and if there was no expert in their midst, all jobs were done by trial and error, with patients learning by imitation. Some patients who had experienced homelessness had excellent wisdom to share with everyone, who were coping with severe poverty in the leprosarium. In addition, doctors treated patients equally based solely on their medical judgment. They distributed extra food only to seriously ill patients. This meant that individuals with a great deal of insight into how to survive were gathered together in the leprosarium.

<4. Self-sufficient organizations for self-defense> were formed because there was both physical and psychological distance from the leprosarium staff, many of whom treated the patients inhumanely. Since the government and the leprosarium did not help them, the patients created self-sufficient organizations such as a youth association, a vigilante corps, and a women's association. These self-governing associations determined salaries based on the difficulty of a given job and were systematically prepared to receive pilgrim patients from outside the leprosarium. Since all residents were assigned jobs, even children, the associations made sure that everyone was doing work suitable for their skills and abilities. The associations created a system to stockpile food and water and made decisions about water use during times of scarcity. If anyone did not follow the rules, a popular leader would make sure they did. During World War II, medical care was even more shorthanded than before because the leprosarium's male doctor went away to war; only the female doctor was left, so patients undertook more of their own care. When traveling actor came to the leprosarium, they listened to patients' requests and solved their issues the following day. In addition, gifts from families living outside the leprosarium helped alleviate patients' poverty. This all shows that patients were self-sufficient and formed organizations within and networks outside the leprosarium.

Patients <5. created a mutualaid system> in order to gather a portion of all payments for compulsory labor and distribute them equally to patients who could not work due to severe sequelae (such as lost eyesight or other critical conditions). They shared food and goods with everyone, especially the weakest among them. If an individual was in poor physical condition and could not work, someone else took their place. The patients were considerate of girls who were ashamed to beg or receive charity and of blind patients who hesitated to receive care from other patients. If an individual had food, he did not eat in front of those without it. On the

http://dx.doi.org/10.5772/65574

other hand, blind patients gave massages to workers as a sign of gratitude for their charity. They all worked to embody the ideal of "helping and loving each other."

Whereas the above categories were most evident before and around World War II, much later the patients <6. fought to compel the government to provide better support>. They initially hesitated to fight the government and work toward abolishing the Leprosy Prevention Law because they feared being forcibly turned out of the leprosaria and returning to the discrimination and homelessness of their hometowns. However, after patients obtained the government's support, their daily lives improved dramatically.

The first two pieces of wisdom (<1. maintaining everyday life through self-sufficiency> and <2. the courage to survive extreme hardship>) showed that the patients had the self-sufficiency and psychological strength to cope with severe poverty. The next three pieces of wisdom (the importance of <3. resourceful people coming together>, <4. self-sufficient organizations for self-defense>, and <5. the creation of a mutualaid system>) demonstrated that, as a group, the patients had extraordinary coping abilities. Using these five categories, Hansen's disease patients survived severe poverty up to and during World War II by forming a skilled community that had subsystems for self-sufficiency, self-defense, and mutualaid. Later, by <6. fighting to induce the government to provide better support>, they showed that their community had the strength and ability to cooperate with other leprosaria groups and fight against the government.

3.2. How we can pass on the wisdom of aging survivors to future generations

The public ostracized Hansen's disease patients, who lived with the stigma, that their illness was due to divine punishment or was a national disgrace. They coped by forming a skilled community and survived severe poverty. Their community had subsystems for self-sufficiency, self-defense, and social security. Their community functioned systemically and was headed by leaders who were hard workers with good morals. The patients were able to cope with severe poverty by themselves, without help from the government or the leprosaria. They maintained their autonomy. This revealed their strength, resilience,and dignity in fulfilling their own responsibilities, even when they suffered. They proved that they could function not only as individuals but also as a group. This is similar to Viktor Emil Frankl's "attitude value," Viktor Emil Frankl [31] developed in his book *Man's Search for Meaning* and as part of his logotherapy, a form of existential analysis. When someone undertakes his life bearing a heavy cross, his destiny is sublimated from mere fact to existential meaning, and he has a "will to find meaning," that is, a need to find meaning in extreme situations. Hansen's disease patients embraced their extreme circumstances as their destiny and found meaning in living with the illness.

In old age, the ego's development task is to find integrity and avoid despair by acquiring wisdom [32]. Wisdom means being detached and transcending concerns about one's own life before death, integrating life experiences, and learning how to pass these lessons on to future generations in spite of decreasing physical functions. Today, the average Hansen's disease survivor in Japan is 82. The survivors have self-confidence and pride in having lived through severe poverty on their own, with little help from the government or medical staff. They are integrating their lives in end-of-life development tasks, and their extreme experiences produce heightened wisdom for posterity.

We think the survivors' knowledge has a universality that can provide suggestions for future difficulties, and is thus valuable for future generations. For example, in the Great East Japan earthquake and tsunami of 2011, many small, depopulated villages were isolated and could not obtain help from outside. The wisdom of the Hansen's disease patients, specifically in forming skilled communities (with subsystems for self-sufficiency, self-defense, and mutua-laid), suggests how isolated villages could heighten their community's resiliency during a crisis. Furthermore, in order to survive in a chaotic society, their wisdom could be used as a business model to create strong teams whose members have good rapport with each other.

What would genuine restoration of survivors' rights look like? Today, their severe experiences are used as a negative example in human rights education. While commendable, we think this is insufficient due to their past and present suffering. We believe that a real restoration of their rights would be utilizing their wisdom to solve future problems. Many people appreciate their knowledge and recognize its value, such as in the spread of logotherapy from Auschwitz. Thus, we hope to pass on not only the negative history of survivors' experiences, but also their insight.

4. Current concerns for survivors and ethical practices for nurses

Above, we discussed the violation of bioethical principles regarding the treatment of patients in leprosaria and the wisdom they gained from surviving extreme situations. In this section, we discuss current concerns for aging survivors and introduce ethical practices for the nurses who work with them.

4.1. Introduction to our research: How will we care for survivors until they all pass away?

Today, the average Hansen's disease survivor in Japan is 82. There were a total of 1718 leprosaria residents in 2015 and 3286 in 2005. Almost all survivors will pass away within the next 10 years. Since former governments attempted to eliminate Hansen's disease by forcing patients to live in lifelong seclusion and to have vasectomies and abortions, most survivors do not have children or grandchildren (except for those who had a spouse and children before entering the leprosaria). As a result, the history of Hansen's disease in Japan will largely end with the death of the last survivor [33] (see **Table 6**).

Survivors are now ensured a comfortable life in their remaining years, thanks to the Act to Accelerate the Resolution of Problems of Hansen's Disease (2008 Act No. 82). As most survivors do not have families, sanatorium staff must care for them. With a decreasing number of survivors, the sanatoriums must now plan how they are going to maintain their facilities. Administrators of the sanatoriums are considering plans to keep them open, such as adding care for the general elderly population. Yet because old leprosaria were built in hard-to-reach places in order to prevent patients from escaping, planning for their future is now difficult.

http://dx.doi.org/10.5772/65574

Maintaining a sanatorium is expensive. For example, one sanatorium is located on a small island 8 km away from the mainland. This island is only accessible by ship. Even if there are only a small number of survivors, it will be necessary to transport medical, welfare, and clerical supplies by ship. In addition, buildings will need to be repaired.

When we think about cost-effectiveness in relation to the country's tight financial situation, one option is eventually combining the 13 sanatoria into one facility. We must balance focusing on the most important goal—ensuring that survivors enjoy a high quality of life until the last one passes away—while also finding a solution to future problems.

This study [33] was conducted at a sanatorium where a new building was constructed for survivors; they were relocated from the old terrace houses that dotted the island to a central nursing home for the elderly. Their former houses were too old and far apart, and nurses could not fully attend to their patients' health. This study was based on interviews conducted with patients 6 months before their expected relocation to the sanatorium and was analyzed using qualitative and inductive methods. The results demonstrated the unique challenges facing Hansen's disease survivors as compared to the general aging population (see **Table 6**).

Category	Sub category
1. The burden of house moving given the age-related decline in mobility and the community's mutual aid abilities	1. Because the survivors are all aging, their ability to help each other is decreasing (e.g., due to the death of colleagues or spouses, and the decreased functioning of the survivors' self-government association, as well as age-related decreases in individual physical function)
	2. Fear of increasing dementia and early death due to changing environment and overwork from relocation
	3. Indeterminate anxiety due to the impossibility of imagining life after housemoving
	4. Bitter memories of past of relocation
	5. The burden of packing and damaging the living environment by carrying household goods
2. The burden of creating new strategies in order to live with multiple and severe sequelae in a new environment	6. Misgivings about being watched while eating among blind survivors who are embarrassed about eating messily
	7. The burden of living with blindness and sensory paralysis, and losing one's mental map of one's environment and having to create a new mental map of a new environment
	8. Suffering the loss of a good residential environment that accommodated ones' sequelae, neuralgia, thermal regulation disorder, and poor vision
	9. Fear of increasing injuries that are liable to be more severe due to the loss of a living environment that had been adapted to accommodate hand and leg sensory disorders and prevent injuries

Category	Sub category
	10. Loneliness and missing a comfortable residential and personal environment, while in one's final abode
	11. The increasing burden of caring for an aging blind spouse
3. The disagreeableness of having one's life disturbed	12. Misgivings about being disturbed from a familiar daily rhythm built over a long time
	13. Concerns about being disturbed due to living near trouble makers
4. Dissatisfaction with the decision-making process managing building structure and room layout	14. Dissatisfaction about inconvenient building structure and room layout
	15. Anger and resignation about not being able to agree on the process of deciding on the structure of the new building
5. Thinking positively to accept relocation and enjoy the new environment	16. There is no anxiety about moving, if our requests are met
	17. All that is necessary is to enjoy a full and independent life before and after relocation

Source: Transferred and modified from Ref. [33].

Table 6. The meaning of relocation for aging Hansen's disease survivors.

The various meanings of the upcoming relocation for aging survivors were integrated into five categories: <1. the burden of relocating from their old home to their new one given the age-related decline in mobility and the community's mutualaid capacities>, <2. the burden of creating new strategies to live with multiple and severe sequelae in a new environment>, <3. the disagreeableness of having one's life disturbed>, <4. dissatisfaction with the decision-making process for managing building structure and room layout>, and <5. thinking positively to accept relocating and enjoy the new environment>.

The most important of these five categories are the first and second. The first one (<1. the burden of relocating from their old home to their new one given the age-related decline in mobility and the community's mutualaid capacities>) shows the various meanings of aging, not just personal aging, but also community aging. The survivors' community is comprised of old men and women because they were not permitted to have children. They have self-confidence due to having survived poverty and forming a skilled community (see Section 3); even after the government began providing further support, the self-governing committee continued to handle daily problems. With the aging and deaths of its members, the community weakened in its ability to cope and exert mobile power. Although residents recognize relocation will likely bring on dementia and hasten death, community aging is as much, if not more, a concern as personal aging.

The second category of concern (<2. the burden of creating new strategies to live with multiple and severe sequelae in a new environment>) contains six subcategories: <6. misgivings about being watched while eating among blind survivors who are embarrassed about eating

http://dx.doi.org/10.5772/65574

messily>, <7. the burden of living with blindness and sensory paralysis, and having to lose the mental map of one's environment and create a new map of an unfamiliar setting>, <8. suffering the loss of a good residential environment that accommodated ones' sequelae, neuralgia, thermal regulation disorder, and poor vision>, <9. fear of increasing injuries due to losing a living environment that had been adapted to accommodate hand and leg sensory disorders and prevent injuries>, <10. missing a comfortable residential and personal environment while in one's final abode>, and <11. the increasing burden of caring for an aging blind spouse>.

The second category shows how important the living environment is for survivors who have multiple severe sequelae. We previously introduced the causes behind repeated wounds and limb loss due to survivors' past physical environment (see Section 2). Today, a survivor's living environment is very important in alleviating sequelae. For example, patients with weak eyesight can recognize changing weather, the transition of the seasons, and the time of day from sunlight entering a window. Survivors with neuralgia alleviate overheating in the summer by using the wind from the sea, without increasing their pain. Thus, over the years, survivors have created suitable living arrangements in order to relieve the pain of their sequelae; relocating means losing their familiarity with their environment.

The second most important aspect of this is the difficulty of redesigning the living environment in order to deal with conflicting or numerous conditions simultaneously, the goal being to mitigate the effects of multiple severe sequelae. For example, some survivors have sensory nerve anesthesia and have lost their eyesight. The general population of blind individuals cannot see, but can recognize items by touch. Hansen's disease survivors who have lost their eyesight cannot recognize items by touching them. They have created a mental map of their environment through trial and error over time, often being wounded in the process, until they are able to move through their homes automatically. By relying on this mental map, survivors know their own position and direction within a space; with the help of a spouse who can see, or listening to the sound of the radio, or feeling a deep sensation of resistance from bumping into the wall, they can easily get around. If they have made a perfect mental map, they can move just as well as if they could see. Thus, forming a mental map is a way of adapting to numerous conditions simultaneously, which helps alleviate multiple severe sequelae. However, when relocating, survivors lose this mental map and must create a new one from scratch.

The third most important aspect of the living environment is protecting the body from external wounds. Since survivors have lost their sense of pain, they do not notice when they have external lesions and their injuries can become severe. Survivors contrive ways to compensate for their sensory nerve paralysis and prevent wounds; relocating to a new environment increases the risk of getting hurt.

The fourth way in which the living environment is important to survivors is that although Hansen's disease causes various symptoms and survivors have multiple grave sequelae, the people influenced most severely by relocating are those who have lost their eyesight. In the past, patients who lost their eyesight could not work; they could not earn a living and fell into poverty. Loss of eyesight was considered one of three major causes of suffering in a patient's life and the other two being notified of one's diagnosis and getting a tracheotomy. In the past, blindness was considered the most serious of the sequelae and continues to be today. In relocating, blind patients must remake their mental map of their environment, but this is difficult because their memory abilities decrease with age.

4.2. What can we learn from this based on a bioethical perspective?

Survivors' living environment can alleviate their multiple overlapping sequelae, but it takes time to adapt a new setting to one's particular needs. This is most challenging for blind patients because changing their environment means they must make an entirely new mental map of their surroundings.

We must now think about how we will care for the survivors until they all pass away. Relocating survivors to one facility may cut costs, but also robs survivors of the setting they have adapted to; this threatens non-maleficence. Survivors need individuals to advocate for their rights, especially as their decision-making abilities and power to acknowledge reality decrease from dementia, making it easy to threaten respect for autonomy. Thus, beneficence is more important now than ever in order to protect the weak.

On the other hand, Japanese medical policy and the medical system must find ways to innovate. As the elderly population grows and the overall population shrinks, the current medical and medical finance systems cannot endure as they are; they must change. Although the law has guaranteed survivors a high quality of life, there may come a day when the nation becomes dissatisfied with shouldering the rising cost of supporting multiple sanatoria. We must seriously consider bringing all remaining patients to one site in order to improve cost-effectiveness.

There is an ethical dilemma in the midst of these circumstances: confronting justice and/or equality, and beneficence and non-maleficence. Unfortunately, we do not yet have a solution. However, we must prevent the public from criticizing and be advocates for survivors. We must find a solution to help sanatoria coexist with society and prosper. If we cannot avoid bringing the remaining patients together at a single facility, we must find methods of decreasing the negative aspects of changing their living environment. We must help society understand the meaning and importance of where survivors live.

4.3. Ethical practices for nurses and end-of-life care

In the section above, we discussed the difficulties of caring for the survivors until they all pass away. In this part, we talk about our efforts to support survivors by providing them with a high quality of life and caring for them until they die peacefully (with support from the Toyota Foundation Research Grant Program 2013). In addition, we discuss the bioethical implications of this practice.

4.3.1. Outline of our projects

At this time, the background of our project is as follows [34]:

a. Aging survivors' deaths cause the community to reduce and collapse over time. There is a negative impact on survivors' physical and mental health as the shrinking of their

http://dx.doi.org/10.5772/65574

community causes them to feel helpless, hopeless, and lonely. It is a feeling comparable to the enforced isolation from their parents that many survivors experienced as children. We must consider how we can alleviate the survivors' sense of loss and prevent growing negative physical, psychological, and social influences.

b. Many survivors do not have any family except for an elderly spouse, so sanatorium staff must care for survivors until they die, instead of their family.

c. Now is our last chance to record the experiences of aging survivors, but many do not have a means of expressing themselves. However, some survivors have produced literary or artistic works.

d. We will lose these storytellers when they pass away. We must consider how we will pass on their experiences.

4.3.2. Expected effects of our practice

As survivors do not have much time left, we looked for an immediate, effective method of working with them to preserve their experiences and communities. We chose life review. With this technique, we can subjectively construct a narrative of their life experiences. It is a process of reweaving one's own life, supported by a good listener (in our case, supportive nurses in the sanatorium), promoting the rediscovery of the meanings in one's life, conducting a reevaluation of life, and gaining a sense of self-consistency.

The expected effects for survivors are as follows: (a) gaining a sense of self-expression through constructing their own narrative; (b) experiencing a cathartic effect, promoting reevaluation of their lives, and developing new meanings; and (c) having a positive effect on survivors by having nurses serve as supportive listeners. As survivors face their own impending deaths, they are acutely aware that the deaths of their comrades have resulted in a shrinking community, thus creating a crisis for survivors. The expected effects for nurses in the sanatorium are as follows: (a) it gives them a chance to exercise their ability to listen attentively and express empathy; (b) it deepens their understanding of the survivors and what they experienced; and (c) it gives them the opportunity to reevaluate their own roles, as well as a sense of responsibility and pride in caring for survivors. These are all nursing practice abilities that are required in end-of-life care. The expected effects for the relationship between survivors and nurses are: (a) a deepening of relationships on both sides and that nurses caring for survivors on their deathbeds will forge a deeper relationship with the patients. The larger point is that supporting survivors' lives and caring for them until they pass away is a nurse's most important duty in a sanatorium. Moreover, nursing practices directly influence survivors' quality of life.

The significance of creating a survivor's life review is: (a) being able to publish the life review in a book, thus passing on survivors' wisdom and experiences to the next generation, and directly contributing to increasing their dignity and satisfaction; (b) since survivors are aging and gradually passing away, this life review book will be the last record of these storytellers and is thus a valuable primary source; and (c) if we continue our qualitative and inductive analysis of these life reviews as raw data, we will find universal meanings in their experiences and will be able to explain their lives plainly and accurately via abstraction.

We conducted life reviews with 17 survivors, with nurses as listeners, and published them in 2015. We included a portion of the life review of Mrs. Hanako Kadowaki at the beginning of this chapter [1]. These 17 life reviews are powerful and fascinating to read; they are useful for thinking about bioethics and about how society cares for people who suffer from discrimination. We can learn both negative history and wisdom from the Hansen's disease survivors. We hope that students of medicine, nursing, pharmacy, and others in the medical field, those who work in Japan's medical administration, and the general population (both young and old people) will read it and learn from it, so as to not allow the suffering of Hansen's disease patients to be repeated in the future. Having worked on these life review projects in the sanatorium, the nurses now have increased self-confidence and pride in their jobs, and can provide even higher quality end-of-life care.

4.3.3. What does it mean to provide high quality end-of-life care as a nurse?

In another part of our research, one of the survivors had been homeless and experienced the hardships of wandering as a pilgrim before entering a leprosarium. This survivor said, "I think my life had a checkered destiny, but I was not miserable. Now, my comrades and I know that the adversity we faced was nothing serious." Thus, the survivor expressed catharsis. In the 17 life reviews we conducted, many survivors showed bitterness toward their past treatment and the policies under the Leprosy Prevention Law, but are now satisfied with and thankful for their current lives. The reasons for their satisfaction and gratitude are: (1) The nation officially apologized for their treatment and provided reparations. Specifically, the Prime Minister, the National Diet, the Minister of Health, Labor, and Welfare, and the Chief Justice of the Supreme Court all issued formal, official apologies. These apologies made the survivors feel relieved. (2) The survivors do not have economic problems, thanks to the new law that aims to resolve all remaining issues of Hansen's disease in Japan. (3) Medical and welfare staff now provide survivors with ample care.

End-of-life care is based on nursing practices. In nursing, we must understand our patients based on holism, practice healthcare based on scientific problem solving, and provide human care based on philosophy. Holism [35] requires looking at the system as a whole, beyond the sum of its parts. We cannot understand the whole only by looking at each element alone; however, through reductionism, we can understand complex phenomena via fields such as biology. In nursing, we understand the notion of "being whole as a person" from Holism. Thus, we grasp at least three points of view and attempt to integrate our understanding. We view our subjects as physical, psychological, social, and spiritual beings. We approach them with consideration for their own life spans. We understand not only pathological and objective diseases, but also the subjective and phenomenological experience of illness. If we want to understand the current thinking and emotions of Hansen's disease survivors, we must grasp their life histories; hence, our project is based on holism. The main characteristics of caring are reciprocity and mutual recognition. According to Jean Watson [36], the purpose of human caring is protecting, maintaining, and enhancing human dignity. In transpersonal caring, both nurses and patients share a spiritual dimension with each other. In other words, when we respect the patient as an irreplaceable, important person, then the nurse and patient can

http://dx.doi.org/10.5772/65574

experience humane and affective interactions. If we have a caring mind, we cannot abuse the patient. Thus, caring prevents deviation from the four principles of bioethics. A peaceful death is an extension of daily care for survivors based on an understanding of them through holism and human caring, which is what our project is founded on.

Japan's political leaders apologized to the survivors, but who heals the survivors' suffering and embodies this apology? High quality daily care by medical and welfare staff in sanatoria directly influences survivors' quality of life. Our project aims to enhance their quality of life and allows them to have a peaceful death. Our nursing practice embodies this national apology and the principles of bioethics.

5. Conclusion and suggestions for the future

1. At present, Hansen's disease is a progressive illness, and new patients live in developing countries. Patients often face discrimination. We must elucidate how stigma is born and adhere to the four bioethical principles in order to prevent discrimination.

2. We think there are three types of people who discriminate: a few support patients, a few persecute them, and the majority do not have correct knowledge of Hansen's disease. If majority agree to persecute patients, they will suffer from greater discrimination. However, if majority support patients, then discrimination lessens. It is important to educate the public and promote awareness of Hansen's disease. Since it is possible that the main people who discriminate have their own psychopathological problems, it is important for patients to understand the real causes of discrimination and how it operates in society, thus putting an end to their challenges and those of people who are prejudiced. There is a lot of bullying and harassment in general and among children. It is important to teach the public that Hansen's disease is not a stigma.

3. The main source of discrimination against Hansen's disease patients is not only based on unpleasant feelings due to the patients' appearance, but also a fear of infection. In the future, should a pandemic of a lethal, drastic infection—such as Ebola hemorrhagic fever—break out, people may ostracize patients due to their fear. We must find ways of protecting against infection yet also avoid discrimination, based on what we have learned from the past negative treatment of Hansen's disease patients.

4. Survivors' experiences living with stigma not only results in suffering, but also wisdom. Like the life-affirming logo therapy that came from a survivor of Auschwitz, the insight gained from living through extreme situations will hearten other suffering individuals and suggest solutions to future problems.

5. In order to solve ethical dilemmas in caring for aging Hansen's disease survivors, we must understand the pathological causes of their symptoms and how they influence daily life for survivors. A pathological understanding is important for solving the ethical dilemmas in treating any disease. When we make a judgment based on the principles of bioethics, we can avoid unnecessary discrimination.

6. In order to alleviate the hardship of oppressed persons, we need a daily nursing practice based on understanding subjects through holism and human caring, thus embodying the principles of bioethics.

Acknowledgements

This work was supported by JSPS KAKENHI Grant Number 15K15797.

Author details

Makiko Kondo[1*], Kazuo Mori[2], Hiroshi Nomura[2], Hanako Kadowaki[3], Makiko Watanabe[3], Akemi Doi[3] and Sayaka Shima[4]

*Address all correspondence to: mkondo@cc.okayama-u.ac.jp

1 Graduate School of Health Sciences, Okayama University, Okayama, Japan

2 National Hansen's Disease Sanatorium Residents' Council, National Sanatorium Oshima-Seisho-en, Takamatsu, Kagawa, Japan

3 National Sanatorium Oshima-Seisho-en, Takamatsu, Kagawa, Japan

4 Yodogawa Christian Hospital, Osaka, Japan

References

[1] Hanako Kadowaki, Mika Seo: Chap. 17 Life review of Hanako Kadowaki, 411–442. In: Makiko Kondo, Oshima Seisho-en (eds.): Life Review of Hansen's Disease Survivors Living in Oshima Seisho-en, Kazamashobo, Tokyo, 2015.

[2] Global Note: Number of incidents of Hansen's disease in the world, compared with countries 2013. http://www.globalnote.jp/post-3941.html

[3] Fujio Ohtani, Hajime Saito, Eiji Nagao, Kunio Murakami (eds.): Hansen' s Disease Medical Science. Tokai University Press, Kanagawa, Japan, 1997.

[4] Makiko Kondo: Chap 21 Appendix pathophysiology of Hansen's disease, 506–519. In: Makiko kondo, Oshima Seisho-en (ed.): Life Review of Hansen's Disease Survivors Living in Oshima Seisho-en, Kazamashobo, Tokyo, 2015.

[5] Keizo Ebina: Biography of Ishidate Morizo. Shinhyoron Publishing Inc. Tokyo; 1997.

http://dx.doi.org/10.5772/65574

[6] Masamichi Goto, Reiko Nogami, Yoshiko Okano, Masaichi Gidoh, Rie Yotsu, Yutaka Ishida, Shinichi Kitajima, Masanori Kai, Norihisa Ishii, Motoaki Ozaki, Kentaro Hatano: Guidelines for treating Hansen's disease in Japan (Third Edition). Japanese Journal of Leprosy 82, 141–182, 2013.

[7] U.S. Department of Health and Human Services, National Institutes of Health: Leprosy (Hansen's disease)—A history of the disease. https://www.niaid.nih.gov/topics/leprosy/Pages/Default.aspx

[8] Shunich Yamamoto: The Japanese history of Hansen's disease. University of Tokyo Press, Tokyo, Japan, 1993.

[9] Ministry of Health, Labor and Welfare: Main event about Hansen's disease July 10, 2016. http://www.mhlw.go.jp/houdou/2003 /01/h0131-5/dekigoto.html

[10] Defense counsel of litigation for national indemnity about Hansen's disease: threatened human rights and their history—the beginning of the policy on cracking down on patients and threatened human rights July 10, 2016. http://www.hansenkokubai.gr.jp/history/history_02.html

[11] Masanao Makino: History of Hansen's disease medicine, 283–298. In: Fujio Ohtani, et al. (ed.) Hansen's disease medical science. Tokai University Press, Kanagawa, Japan, 1997.

[12] Taeko Miura: Chap. 21 Appendix history of Hansen's disease, 520–531 pp. In: Makiko kondo, Oshima Seisho-en (eds.): Life Review of Hansen's Disease Survivors Living in Oshima Seisho-en, Kazamashobo, Tokyo, 2015.

[13] Oshima Seisho-en: 50 years of History at the National Sanatorium in Oshima Seisho-en. Oshima Seisho-en, Kagawa, Japan, 1960.

[14] National Hansen's Disease Sanatorium Residents' Council: Records of Fighting by National Hansen's Disease Sanatorium Residents' Council (reprinted). Ikkosha, 2002.

[15] Ehime prefectural government office: All body "The Leprosy Prevention Law." July 10, 2016. http://www.pref.ehime.jp/h25500/4404/shinpou.html

[16] Study group welfare problem (ed.): History of a national sanatorium (leprosy), Ministry of Health and Welfare, Tokyo, Japan, 1975.

[17] Ministry of Health, Labor and Welfare: Act to accelerate the resolution of problems of Hansen's disease July 10, 2016 (2008 Act No. 82). http://www.mhlw.go.jp/topics/bukyoku/kenkou/hansen/kokuji/dl/9.pdf

[18] Japan Law Foundation: Final reports—verification meeting about Hansen's disease problems, 2005. http://www.mhlw.go.jp/topics/bukyoku/kenkou/hansen/kanren/dl/4d1.pdf

[19] Mikako Yamaberi, Matsuko Kondo, Kazue Ishikawa, Yoshiko Amano, Makiko Kondo: Causes of and coping strategies for repeated wounds in patients with Hansen's Disease

until promin treatment. Journal of Japanese Society of Nursing Research, 38(1), 59–71, 2015.

[20] United States Holocaust Memorial Museum: United States Holocaust Memorial Museum note—Nuremberg Code July 10, 2016. https://www.ushmm.org/information/exhibitions/online-features/special-focus/doctors-trial/nuremberg-code

[21] World Medical Association: WMA Declaration of Helsinki—ethical principles for medical research involving human subjects. http://www.wma.net/en/30publications/10policies/b3/

[22] The National Commission for the Protection of Human Subjects of Biomedical and Behavioral Research: Belmont Report—Ethical Principles and Guidelines for the Protection of Human Subjects of Research. https://videocast.nih.gov/pdf/ohrp_appendix_belmont_report_vol_2.pdf

[23] World Medical Association: WMA Declaration of Lisbon on the Rights of the Patient July 10, 2016. http://www.wma.net/en/30publications/10policies/l4/

[24] J. Gallagher, S. Gorovitz, R.J. Levine (ed.): Biomedical Research Ethics: Updating International Guidelines: A Consultation. CIOMS (The Council for International Organizations of Medical Sciences) Publication, Geneva, Switzerland, 2000.

[25] Toru Sekine: Genealogy of Medical Ethics—The Ancestors' Wisdom for Treating Patients Kindly (revised ed.). Hokujyu Press, Japan, 2008.

[26] Tom L. Beauchamp, James F. Childress: Principles of Biomedical Ethics (Third Edition), Oxford University Press, Oxford, UK, 1989.

[27] Michio Miyasaka: Methodology of Biomedical Ethics—A Principle-based Approach, Procedure-based Approach, and Narrative-based Approach. Igaku-Shoin, Tokyo, Japan, 2011.

[28] Center for Medical Education & Information: The four principles of bioethics July 10, 2016. http://www.c-mei.jp/index.html

[29] Japanese Nursing Association: Information for nursing practices—ethical principles July 10, 2016. http://www.nurse.or.jp/rinri/basis/rule/

[30] Takahiro Tanikawa, Hideko Miyawaki, HitomiShinjyo, Yoshiko Amamo, Makiko Kondo: The meaning of experience in the everyday poverty for the oldest Hansen's disease survivors—through the narratives of residents admitted around World War II into a leprosarium in the Seto Inland sea. Japanese Journal of Leprosy, 84(1), 37–50, 2015.

[31] Viktor Emil Frankl: Man's Search for Meaning. Beacon Press, Boston, USA, 2006.

[32] Erik E. Erikson: Identity and the Life Cycle. WW Norton & Co. Inc., New York, USA, 1994.

[33] Mikuni Oyabu, Hitomi Yamada, Rie Nishimoto, Maki Sanada, Kyoko Tsukimori, Makiko Kondo: The meaning of relocation among aging Hansen's disease survivors in Japan. International Journal of Nursing & Clinical Practices, 3, 172, 2016. doi: 10.15344/2394-4978/2016/172

[34] Makiko Kondo: Chap. 1 Significance of the project, 2–13 pp. In: Makiko Kondo, Oshima Seisho-en (eds.): Life Review of Hansen's Disease Survivors Living in Oshima Seisho-en, Kazamashobo, Tokyo, 2015.

[35] Keiich Noe: Holism. Philosophy Dictionary, Iwanami Shoten, Japan, 1998.

[36] Jean Watson: Human Caring Science: A Theory of Nursing (Second Edition.), Jones & Bartlett Pub, Chicago, USA, 2011.

Chapter 6

Rethinking the Postwar Period in Relation to Lives Not Worth Living

José-Antonio Santos

Additional information is available at the end of the chapter

http://dx.doi.org/10.5772/66394

Abstract

This research will focus on the postwar period in relation to lives not worth living. This chapter is divided into five sections. The first section is a short introduction to the overall topic. The second part discusses the legal and philosophical language, post-Second World War, in relation to the psychically and mentally ill. This raises the question of whether or not philosophy can be made after 1945 without looking at Auschwitz. Adorno's categorical imperative: Auschwitz can never ever be repeated, gains prominence in the way of making and proceeding in philosophy and in law this does not include just the Holocaust but also one of the most forgotten groups: the severely mentally and/or physically disabled. The policy of oblivion was practised much quicker than with other human categories. The paradigm of human rights changes substantially immediately after the Second World War The establishment of individual responsibilities for the committed atrocities will be carried out by means of the Nuremberg's Trials. The third section focuses on the Nuremberg War Crimes Trial, USA vs. Karl Brandt et al. The fourth section analyzes the concepts of post-Auschwitz memory and memory of oblivion. Recovering post-Auschwitz memory implies recreating to the thought process after 1945. Finally, the fifth section draws some conclusions and indicates some further areas for research.

Keywords: disability, "euthanasia," life unworthy of life, memory, postwar

1. Introduction

The Nazis rise to power in 1933 marked the start of a historic stage featured by its moral decay. This decline which had its breeding ground in Romanticism, framed in the context of modernity, continued some time later during the Weimar Republic, although within a different reference system. The consequences were, at the linguistic level, the creation of an ethical, legal and philosophical language able to substantiate one of the greatest and more perfect machineries of extermination of humanity, with a clear organic component based more on feelings than on reason.

Analyzing this problem requires a prescriptive approach. The reason is that the Second World War represents a more or less generalized fracture in the history of contemporary thought. It is particularly evident in philosophy and, above all, in law. In this context, this chapter will look at the rule of memory as an instrument to analyze events already occurred, as experience for present generations and, in particular, for the future.

With this scenario, it is appropriate to take stock of what happened and gradually rebuild the postwar period, while paying attention to the mentally and/or physically handicapped during the Nazi era, groups which underwent such a quick policy of oblivion. This group serves almost as an excuse for rethinking the memory, from a hermeneutical perspective, about a group of people invisible to society and whom history largely ignored. With regard to this point, two aspects are analyzed: the first, concerning the state of the art after 1945, which meant a substantial change in the legal and philosophical thinking. Its implementation was done by means of a particular denazification that aimed at countering the atrocities committed during National Socialism. The second, however, refers to the post-Auschwitz memory, which involves reflecting with a reasonable use of ethical, philosophical and legal language; the aim of which is not wallowing in the Germans' guilt, but helping to explain this starting point of how to reflect and maintain a responsible consciousness alive among the citizenry. In this regard, from a retrospective vision, it is observed how the meaning of the words has sometimes been gradually changed for the sake of spurious interests beyond a reasonable frame.

The philosophy of the memory—beyond partisan utilization—naturally avoids confusing the political commitment with the intellectual rigor, since the memory is objectionable when used as a political weapon. The memory of oblivion cannot be but selective and endowed with a halo of impartiality, but it should not be confused with an imposed oblivion that would justify amnesty. Rethinking the past is to confront the present with other eyes: to think it over again. That is, with commitment and without complacency, which make the individual not turn his back on others while carrying out such acts. Remembering entails a task of learning, key in this sense how to remember, this is, how what is learned must be understood. In this context, there arises an interesting aspect that determines the study: the more the victims suffered, the more they want to disregard the suffering of others and the generations to come.

2. Transition to a New Era

Before the end of the Second World War, the leaders of the United States, Russia and the United Kingdom projected, in the 1943 Teheran Conference, how to create an organization for peace once war was over. During this time, there was a feeling that injustice prevailed and that what had happened should never happen again. This way of thinking leads to an

http://dx.doi.org/10.5772/66394

unprecedented change of scenery, which greatly affects the way of understanding philosophy and law. After the crimes against humanity perpetrated by Nazism, by means of the extermination camps, the maxim of Hegel, "the real is the rational and the rational is the real" [1], lost part of its meaning. The unreasonableness of National Socialism had overcome any type of reason. By contrast, Adorno's words were to gain prominence: "The feeling which after Auschwitz resists every assertion of positivity of existence as sanctimonious prattle, as injustice to the victims; which is reluctant to squeeze any meaning, be it ever so washed-out, out of their fate, has its objective moment after events which condemn the construction of a meaning of immanence, which radiates from an affirmatively posited transcendence, to a mockery" [2]. As a categorical imperative (Auschwitz can never ever be repeated) [3], Adorno could not have foreseen that genocide would also happen in Cambodia, Africa or Bosnia, highlighting the extreme weakness that features memory. So much so that, according to Mate, there is a relationship between the current oblivion and the amnesia that characterized the historical logic that led to catastrophe, which is why history can be repeated [4]. The human being should consider who he was after what happened and, above all, who he wanted to be. If cultural, rather than moral, standards had prevailed in the Weimar Republic, the opposite happened postwar: a culture of exclusion and alienation had prevailed during National Socialism, so the postwar period was built on the basis of awareness and a nuanced inclusion.

The year 1945 signalled a before and after in the history of contemporary legal and philosophical thought. On March 30, Hitler took his own life and, with it, the hopes of his supporters began to fade. However, the United States put an end to the Second World War by dropping atomic bombs on Hiroshima and Nagasaki in August of 1945, thus terminating the conflict and giving victory to the Allies over the Axis Powers. The fact that North America returned democracy weighed on the Germans' conscience. In practice, though, Germany remained occupied and divided by the four victorious powers for more than 40 years: the western part, run by France, Great Britain and the United States, and the eastern part, by the former Soviet Union.

The welfare state had begun to take shape during National Socialism, so that the progress from a totalitarian state to a welfare state occurred gradually. Nonetheless, much of Germany was devastated and criminal acts occurred more frequently than expected. The postwar period brought about strong institutions designed from the ground up, but sometimes they did not have the character of true universality as was originally intended. Having left behind semantic transformations, it was time to speak of "the grammar of rights" [5]. As was clearly reflected in the preamble to the Charter of the United Nations: "We the peoples of the United Nations determined to save succeeding generations from the scourge of war, which twice in our lifetime has brought untold sorrow to mankind, and to reaffirm faith in fundamental human rights, in the dignity and worth of the human person, in the equal rights of men and women and of nations large and small, and to establish conditions under which justice and respect for the obligations arising from treaties and other sources of international law can be maintained, and to promote social progress and better standards of life in larger freedom" [6].

Along similar lines, the Universal Declaration of Human Rights of 1948 was, for 51 countries, a fresh start, which emerged—according to Orrego—at a time of radical self-awareness of humanity regarding their ability both to exceed in evil, as to make justice prevail through the use of force. In view of the Holocaust, the attraction of relativism and scepticism in relation to the dignity of the human person paled for a moment [7]. Human dignity stood out as unavailable, while at the same time an ambivalent result was forged: there was no need to argue on the basis of human rights, because they were already included in writing in an agreed document. They appear also as a historical notion of contingent: the institution of human rights exists, but it could cease to exist as well.

Despite going back to the beginning, terms such as eugenics and "euthanasia" were already in the minds of German physicians, as well as in certain circles of power. According to Gadamer, language served as a means to perform the agreement of the partners and the consensus on the matter [8]. The readings of Nietzsche plus earlier and later authors had left a residue for the configuration of a really pejorative kind of language: *Ballasexistenzen, lebensunwertes Leben, Nebenmenschen, Untermenschen*, etc. In short, the seriously mentally and/or physically disabled were reduced to nothing. Other groups also suffered the consequences of the establishment of racial hygiene, although with varying degrees of a greater recognition: Catholics, communists, homosexuals, social democrats, etc. Medical professionalism and consent did not exist as concepts yet. It took this several decades for this situation to change and possibly, to the other extreme. The mentally and physically ill never granted consent to their sterilization or extermination; however, some of them knew what their fate would be. The mentally and/or physically handicapped ended up becoming invisible human beings in society for various reasons: On the one hand, their families did not want the facts to be known because of fear of rejection, whereas on the other hand, they were not supported by institutions as were other groups. They were quickly pushed into the background, in comparison to other groups, such as the Jews, playing a minor role in the collective memory, comparable only to heterogeneous asocial groups, which could even be listed under different categories given the difficulty of classifying them. The lack of political interest that the asocial suffered is proven by the fact that they were later excluded from the measures of compensation granted to other groups which were subjected to the policy of extermination. Moreover, according to Muñoz Conde, some politicians who had been with them side by side in the concentration camps expressly rejected that also the asocial were compensated, because if they were compensated the same way as others, they argued, this would confirm a perverse thesis of National Socialism: equate political opposition with criminals and the antisocial [9]. Few remember their faces or their names. Similar fate befell the mentally and/or physically handicapped. Chasing after the sick and, in general, the "misfits" would become a constant of the Nazi regime, as well as the obsession with children, since they would be those who would determine the purity of blood and race.

After Auschwitz, came a new era with an entirely different philosophy and law; in short, an unprecedented awakening of the use of memory. The old German law—as noted by Stolleis—was replaced by a new one: in particular, striking symbols or symbolic figures of the old regime were immediately eliminated [10]. Gone was the notion that argued that the Nazi legal thought came from the "spirit of positive law" [11]. According to Neumann, the official

http://dx.doi.org/10.5772/66394

legal ideology of National Socialism is methodologically closer to "iusnaturalistic" thought of the traditional systems of natural law that in positivist conception [12]. Natural law was used ambiguously; during National Socialism, as a natural law of the race, while during the Second World War as *neo-iusnaturalism* conveyed through the nature of things. It is a fact that the existence of not a few supporters of natural law in conjunction with the nature of things, circumstance that loomed the terminological and ideological variety which these concepts could entail. Especially in the period, the post-Second World War, the focus was on establishing *objectives* values and an attempt at ranking them, in order to overcome the crisis of values in Nazi Germany.

The differences between the first and the second postwar periods were more than palpable. Bessel highlights how the shock of the cataclysm of 1945, characterized by extreme violence, the scale of the German defeat, occupation and division of the country, and the fight often desperate for individual survival, made possible a transition that is very different from that following the First World War [13]. This was a complex and difficult to manage transition, to which they had to articulate a system which tried to counter the perversion which occurred years before. A system that was the cornerstone of a particular asymmetrical settling of scores, the Nuremberg Trials, asymmetrical because the horrors suffered during the Nazi regime ended with a handful of executions and convictions of some important Nazis which silenced some voices. Others, though, had already lost their lives to suicide or had managed to escape justice. A series of denazification processes were initiated on November 20, 1945, and extended up to April 14, 1949.

3. Social Imaginary, Politics and Law in Nuremberg

On this occasion, focus is on the Nuremberg War Crimes Trial, USA vs. Karl Brandt et al., despite the fact that important processes were conducted by national courts, too. For example, in the so-called Hadamar Trial in Frankfurt in 1947, a court judged several employees of that centre for the death of 10,000 patients within its facilities between 1940 and 1945. The Hadamar Trial focused on doctors, nurses and both administrative and technical staff. One consequence was that the official in the area of Wiesbaden, Fritz Bernotat, head of psychiatric institutions in Hesse, was never really prosecuted [14].

On December 9, 1946, The Doctors' Trial began for war crimes and crimes against humanity, which would culminate on August 20, 1947. An international military court, composed of three judges representative of each victorious power, issued a conviction against a total of 23 people, 22 men and 1 woman, of which 16 pleaded guilty. Once the entire process was over, the result in some cases was less onerous than expected. Among those accused of committing war crimes and crimes against humanity and for belonging to a criminal organization, only seven ended up sentenced to die upon the gallows: Viktor Brack, Karl Brandt, Rudolf Brandt, Karl Gebhardt, Waldemar Hoven, Joachim Mrugowsky and Wolfram Sievers, whereas Fritz Fischer and Karl Genzken were sentenced to life imprisonment. A second group was found guilty of war crimes and crimes against humanity, and the group was sentenced to the following penalties: Siegfried Handloser, Gerhard Rose and Oskar Schröder, life imprisonment; Hermann Becker–Freyseng and Hertha Oberheuser, 20 years in prison; lastly, Wilhelm Beiglböck, 15 years. Third, Helmut Poppendick was sentenced to 10 years in prison for belonging to a criminal organization. Finally, Kurt Blome, Adolf Pokorny, Hans Wolfgang Romberg, Paul Rostock, Siegfried Ruff, Konrad Schäfer and Georg August Weltz were acquitted [15, 16]. All of them did their utmost to improve race to the ultimate extreme.

The prosecution strategy was to prove "euthanasia" to be a war crime, so that military justice had jurisdiction over the case. In general, it was intended to attack medicine as ineffective, unscientific and monumentally destructive. The defendants and their lawyers legitimized "euthanasia" to understand it as a relief from the suffering of the severely disabled and incurably ill whose lives were unworthy of living. At the same time, Karl Brandt and Viktor Brack claimed that this "euthanasia" was not really different from practices defended in other countries. To do this, they cited a series of literary justifications of "euthanasia," including essays of the Austrian Jew and socialist eugenicist Julius Tandler, and excerpts from the work *Man, the Unknown* by Alexis Carrel. Karl Brandt denied having general power on medical issues, which would be extended to euthanasia and genocide, and having been involved or having had knowledge of the human experiments. Thus, he made a distinction between "euthanasia" as ethically and medically justified in individual cases, and the mass murder of psychiatric patients. Even Karl Brandt asked to use, at the trial, the films *Ich klage an*, *Life Unworthy of Life* and *Existence Without Life*, to prove to the Court that "euthanasia" was ethical and human [17].

The various convictions had an exemplary character, in spite of the fact that those imprisoned were released before having completed their sentences in full. According to Zolo, the start of this international criminal justice appeared uncertain and controversial from several points of view: the independence and impartiality of the courts, in particular of the Attorneys General, the respect for the rights of habeas corpus for defendants, the quality of the penalties inflicted on the condemned, its purpose and preventive efficacy [18]. The deaths of those sentenced could not be a private issue, so they were hanged for their crimes, although not as much publicity was given to this as would have been desirable for certain sections of the population. Nobody remembered or gave enough thought to what had happened.

In the above-mentioned process, a number of people who had made possible the use of a type of medicine of the most advanced in the world were prosecuted. Normal individuals with ordinary lives, who had trained well in college, studied art, science and ethics of medicine, and who had sworn the Hippocratic Oath at the end of their studies. So, what went wrong and what had failed? In this context, Hassenfeld focuses more on psychiatrists, being difficult to establish what went wrong [19]. Lights and shadows cast on the causes of the quick accommodation of the physicians to the programs of racial hygiene to experiment with and, ultimately, annihilate human beings whose lives had no value for the Nazi regime. Unfortunately, it is known how little publicity that those sentences were given, as with the other legal decisions that were issued in Nuremberg.

The population rejected the eugenic and "euthanasic" practices, but the response was more dramatic than expected: silence prevailed. The reason for this attitude is far from clear within

http://dx.doi.org/10.5772/66394

the Academy. The Adenauer government was influenced by that attitude with a more anti-Communist than anti-Nazi policy, based on which they were not that interested in opening proceedings against former Nazis [20]. Klemperer lucidly shows his displeasure labelling it as a "deadly disease" to the so-called denazification. As he states: "I hope, and indeed believe, that this dreadful word will only have a short life; it will fade away and lead no more than a historical existence as soon as it has performed its current duty." A word that will someday fall into extinction, as if the situation that it would have ended, will no longer exist. This will still last a while, "because it isn't only Nazi actions that have to vanish, but also the Nazi mindset, the typical Nazi way of thinking and its breeding-ground: the language of Nazism" [21]. Unfortunately, it has yet to disappear.

As a result of the findings of the judges arising from that process, the Nuremberg Code—on ethics in experimentation with human beings—was promulgated in 1946, as a starting point for the development of the limits in medical research. Among other matters, voluntary consent of the patient raised as absolutely essential; research based on previous animal tests, together with a knowledge of the natural history of disease that may promise results to justify their implementation; cautions related with risk taking; measures that were appropriate to safeguard any kind of prejudice (injuries, death, etc.) that might be caused; the possibility of ending the experiment by both patient and physician, and the training of the health workers involved [22].

The preparation of various agreed documents such as the Declaration of Helsinki [23] of 1964 by the World Medical Association (it would be subjected to successive later versions: 1975, 1983, 1989, 1996, 1999 and 2000) did not impede the implementation of eugenic policies with varying degrees in different countries of the world. It is easy to remember cases of "early euthanasia" as the known process of thalidomide in Liege 1962, in which a mother and a doctor who had killed a child born without arms were acquitted. A British doctor was also acquitted in the early 1980s; the physician helped a child with severe Down syndrome to "sleep" through the use of tranquilizers and sedatives. Meanwhile, another doctor was convicted of attempted murder—in 1970 in Hamburg—for throwing away, along with the placenta, a premature child to whom he did not give the possibility of survival [24].

It is nonetheless surprising that medical experiments were still conducted, without too many barriers, after all that has happened. The ruling Buck v. Bell remained the best defense of the legally established sterilization, even in the inmediate aftermath of the Holocaust [25]. Until the 1970s, State measures of sterilization were conducted as the famous Tuskegee experiment developed from 1932 to 1972 in Alabama. With this experiment, according to Lombardo and Dorr, eugenics reinforces and updates the "racial medicine" of the nineteenth century, establishing it firmly on modern and scientific grounds [26]. The Belmont Report was issued, which described the three principles of bioethics (respect for persons, beneficence and justice) that were then considered, as a result of that tragic event, among other circumstances. Later, the principle of nonmaleficence (*primum non nocere*) was added to the three [27].

4. Post-Auschwitz Memory and Memory of Oblivion

The survivors of the concentration camps and the sanatoriums were released, after the Nazi regime ended. They would become the living image of memory and testimony. If nothing is done to keep their memory alive, their death will bring about oblivion. Against this background, the task of thinking post-Auschwitz becomes fruitful and necessary in order to build a discourse that would prevent the Nazi injustice, the pointlessness of that event [28]. Contemporary thinking conditioned by Auschwitz which has an "epochal value," "therefore" there is a before and an after, also for philosophy" [29]. Nevertheless, once the trauma post-Auschwitz seemed to be overcome, the following step is to rethink and to reinterpret a memory of oblivion, which supplies its effect when it is accompanied—as Ricoeur says—of a "duty of memory" consisting of a "duty to not forget" [30].

After the Second World War, the person is taken as a reference in the field of law, later in bioethics, although the passing of the years has diluted that anthropological conception underlying the legislation and jurisprudence of the fifties to seventies in Germany. There were mainly two attitudes about how to rethink memory: those people who wanted to forget and that of those who tried to remember. If you opt for the latter, it is commendable to understand that, according to Mate, Auschwitz could not be planned in all its horror, but it did happen. And when the unexpected occurs, it becomes a reference for thinking, that is, something that makes one wonder. Thinking after Auschwitz means taking us back to that point of negativity when thinking of the problems of our time. With this, we want to express that "memory is the beginning of thought. There has been a substantial change in the social value of memory and its contents. In fact, memory value rises higher. It is a global phenomenon" [31].

Such a major philosophical process highlights the interest in revitalizing the post-Auschwitz memory—as already indicated at the beginning of the work—in that mental faculty of retaining and recalling past events, located in a space-time context after the Second World War. This concept should not be assimilated to historical memory, despite the relatively recent development of the second concept. The reasons vary and are of different degrees: on the one hand, memory should not be spiteful or vengeful. Sometimes, as Ricoeur states in relation to memory, it resembles more of a "competition among ideologies" than a serene debate on rational arguments: "It is always the other who joins in ideology" [32]. On the contrary, you should try to be fair, at least, from the starting point, because the point of arrival can never be so, since memory is by definition *partial* and *subjective*. Obviously, it is extremely difficult to talk about absolute impartiality and objectivity both in legal and philosophical knowledge.

By its own idiosyncrasies, the notions of history and memory tend not to reconcile to the extent that history may be the destruction of memory [33]. History involves the destruction of memory, because on several occasions, the elements used in the memory as retention or memories do not exceed the methodological guidelines used by historians. According to Gustavo Bueno, history, as a scientific discipline, is not the product of memory, nor has it anything to do with the memory more than chemistry or mathematics may have, either. History is not simply a reminder of the past. History is an interpretation or reconstruction of the relics (which remain in the present) and an arrangement of these relics. Thus, history is the work of understanding, not of memory [34]. The intent does not imply *making history* as pursued

http://dx.doi.org/10.5772/66394

by the historian, but rather *making up memories* [35] as the philosopher poses. This is normally made collectively or individually by definition, but all in all the first may consist of a kind of reasonable individual memories. So much so that—according to Mate—the substantive sense of memory is not the time to reminisce but to bring to the present, to recognize the validity of the marginalized past, the past of the losers and not only the losers but also the forgotten [36]. A task of such magnitude acquires special relevance in all the tradition that raises memorizing to *ars memoriae* [37] and interpretation to *ars interpretandi*. Both text and context are more than necessary (for the need to remember that combines a range of *reasonable individual memories*) in order to build a complete memory.

Remembering implies approaching the past historical context, in order to interpret its concepts, to detect errors and not repeat them in the future. Heroism, patriotism, and uniforms were three hallmarks during National Socialism. Germany, always in search of its identity, tried to find the one that best fulfilled its role, postwar: heroism. If society is not aware of the positive task accomplished by those unsung heroes for humanity, efforts by groups like the White Rose (*die weiße Rose*), the Kreisau Circle (*Kreisauer Kreis*) or the Solf Circle (*Solf–Kreis*) will have proved barren. The children of the post-Auschwitz generation will always throw in the faces of their parents that they were not strong enough. Arendt describes the attitude of certain German youth expressing feelings of guilt as cheap sentimentality, while they carry undeterred the burden of their parents' guilt [38]. Instead, Arthur Kaufmann postulated a resistance against apathy of the heart, that is, resistance against indifference and resignation which meant permanent suicide [39]. They wanted to see that it was not the time for heroes, but for the acquiescent. The concept of heroism was discredited and relegated to the background, because heroism without uniform lacked relevance. In fact, "Nazism didn't recognize any kind of decent, real heroism. It thereby perverted the whole notion and brought it into disrepute" [40]. Heroism was replaced by symbolism, officers in their uniforms. A uniform created by the only party that made their power absolute with the help of the media. A particular symbolism was reflected in the different films and documentaries about the sanatoriums for the mentally handicapped: on the one hand, the mentally ill with their typical clothes and the images of their disabilities; on the other hand, images of health professionals with their pristine white coats, patiently dealing with human beings labelled misfits and sometimes violent, with insignificant lives.

In the legal and political language of the Third Reich, the concept of patriotism suffered a radical turn, when it was connected with a particular nature, race and culture: the Arian. Patriotism was changed to slavery an interpretation of the exclusionary, ethnic and imperialist homeland emphasizing the mythical past features, while rebuilding history mixed with legend to rouse the masses. Goebbels stated that the first commandment, every Nazi should know: your fatherland is Germany. Love it more than anything, and more in deed than in word [41]. On the contrary, according to Mate, a more feasible path passed by understanding the subject of History not thinking of the happiness of our heirs, but rather of the sufferings of our ancestors [42]; less patriotism and more accountability.

All that paraphernalia caused an ambivalent reaction of measured admiration and outright rejection in a number of future intellectuals. It is possible to speak of three groups within these: those who recalled the past, those who prefer to forget the past and those who were trying to live in the present thinking of the future; while it is true that these three groups did not function as watertight compartments, but rather some of them could "infect" others. However, as a basis for arguing, it is easy to end up using stereotypes, for example, the young against the old. Historic responsibility of later generations has been marked by a past they barely knew.

The value given to stability contributed, according to Bessel, to reconfigure the successful conservative culture of the 1950s in Western Germany and to undermine the policy of continuous alterations and transformation driven by the Government of the Democratic Republic. Seeking security was a key factor in Germany's postwar culture [43]. In the 1950s, the injustices of criminals were considered as justified, as part of the popular clamour, which led to a second set of proceedings aimed at a less unpopular outcome than the first. The executioners became victims. The strategy was to discredit the Nuremberg Trials as the victors' justice, even as the Jewish and the winning leftist conspiracy in the form of a military tribunal with the ethos of an international tribunal [44]. Years later, the prescription of the crime was discussed and even certain sectors called for a general amnesty. There was not a real and serious pursuit of Nazi criminals. The consequences of the T4 action and eugenic policies were minimized. Not wanting to recall showed a biased view of the facts, an amnesia caused by a deliberate interest in forgetting. Wiesel takes stock of that tragic situation several decades later: "When I think about the Nazi doctors, the medical executioners, I lose hope. To find it again, I think about the others, the victim–doctors; I see again their burning gazes, their ashen faces" [45].

Different processes carried out against the Nazi criminals aroused in the younger generations, who had known very little about National Socialism, an eagerness to learn more about the events that occurred. The effort to make the Nazis palatable suffered a new setback with the arrest of Eichmann in Buenos Aires on May 11, 1960, whose mission was to direct the program that carried out Hitler's Final Solution. The declaration of Eichmann's guilt by the Court meant a provisional end point, being sentenced to death and executed on June 1, 1962; a process that had begun a year earlier. It marked a before and after by the magnitude of the crimes and his denial of them, arguing that he obeyed rules of war. A stage of more Marxist content started with a Germany divided by the Wall, which was unable to see at base level. This was a pointless confrontation. The good times were not necessarily those of others or our own, it would rather be preferable to talk about time with the others. Adorno saw the paradigm shift clearly: the requirement that Auschwitz would never happen again is above all a matter of education; hence, it should be taken seriously with an intention that is not alien to philosophy [46].

Arendt would take over to deepen that barbarism, after having described the last moments in the life of Eichmann: "It was as though in those last minutes he was summing up the lesson that this long course in human wickedness had taught us–the lesson of the fearsome, word–and–thought–defying *banality of evil*" [47]. Later, Habermas would further comment on that tragic event, although from a different perspective: in Auschwitz "it touched a deep layer of solidarity among all who have a human face. Until then—in spite of all the quasi-natural brutalities of world history—we had simply taken the integrity of this

http://dx.doi.org/10.5772/66394

deep layer for granted. At that point a bond of naiveté was torn to shreds—a naiveté from which unquestioned traditions drew their authority, a naiveté that as such had nourished historical continuities. Auschwitz altered the conditions for the continuation of historical life contexts—and not only in Germany" [48]. Its uniqueness was primarily focused on the Jewish people, but also serves as a reminder for all victims, especially the forgotten. That fact can be understood as a particular event and, above all, singular, because with Auschwitz—according to Mate—"humanity reaches a hitherto unknown degree of inhumanity" [49]. Auschwitz epitomises the paradigm of the Holocaust, an inexplicable case that is incomparable. Later there would be other Holocausts. Rethinking memory involves doing so from serenity and prudence, but firmly. This type of memory intends to rethink not only the law, philosophy and ethics but also to a great extent the language.

5. Conclusions

Globalization brought a perversion of the language, in its interest for making it deliberately universalist, to understand well what has been learned, while perhaps it did not globalize the ability to understand. Language is key and altering it can only lead to catastrophe. It is necessary to avoid verbal hypocrisy and semantic pollution, so as not to fall into an irrational pointless situation. Language has not changed much from that period to now, contexts are different but trends are similar [50].

Just as bioethical and biolegal issues end in a kind of covenant, making that well-intentioned thoughts could lead to the error of a mistaken assessment of the course of action. It is of the greatest importance that this stance be committed to and be truthful to reality. The past should be assessed while taking care that it is a sensible past aligned with political equity of all human beings. Nihilist postliberalism of our epoch notes its indifference towards others. The point was not to live for others, but not to turn their back on others. It is necessary to put an end to this type of Nihilism, which is even more disruptive than scepticism itself. Within this context, the point is not censoring every research about evolution or genetics, but to call for due caution when making irreversible decisions of legal, medical or moral type on others.

The Nazi genocide is always remembered as the paradigm of the Holocaust, although there were also indiscriminate killings which cannot be labelled as genocide since it did not affect a specific ethnic group as they were patients with serious physical and/or mental illness. On this basis, it is appropriate to act responsibly in taking further action, in looking back and reflecting, ultimately coming to our present, burdened with our past, a two-way process that will help in understanding the past charges of our present.

By exposing this issue, I sought to highlight a better common legacy for future generations, rethinking memory. Thus, it is crucial to decide to have a future with or without memory. If the first option is chosen, it is important to note that the advancement of such memory will be more learned than lived. When all survivors of Nazism disappear, only the learned memory will remain. Memory inevitably presents a selective character, which is always featured by a dreamt memory component. Times of yesterday make the past mix with the present, because without remembering the past we will have committed the excesses of wanting the present too much. A memory from oblivion should be told not only from the perspective of the vanquished and remembered but also from the losers and the forgotten.

Acknowledgements

This is the English modified version of the chapter 4 of my book 'Los olvidados del nacionalsocialismo. Repensar la memoria'. Madrid: Centro de Estudios Políticos y Constitucionales; 2014. The author is grateful to the publisher permission to publish this work. The chapter was translated by María-José Pastor and revised by the author.

Author details

José-Antonio Santos

*Address all correspondence to: joseantonio.santos@urjc.es

Faculty of Law and Social Sciences, Rey Juan Carlos University, Madrid, Spain

References

[1] Hegel G.W.F. Werke (*Works*). Vol. 7. Moldenhauer E., Michel K.M., editors. Frankfurt am Main: Suhrkamp; 1979. p. 24.

[2] Adorno T.W. Negative Dialektik (*Negative Dialectics*). Frankfurt am Main: Suhrkamp; 1966. 352 p.

[3] Adorno T.W. Erziehung nach Auschwitz (*Education after Auschwitz*) (1966). In: Tiedemann R, editor. Gesammelte Schriften. Vol. 10, part 2. Frankfurt am Main: Suhrkamp; 1966. p. 24.

[4] Mate R. Por los campos de exterminio (*For the Death Camps*). Barcelona: Anthropos; 2003. p. 168, 51.

[5] Orrego C. La 'gramática de los derechos' y el concepto de derechos humanos en John Finnis (*The 'Grammar of Rights' and the Concept of Human Rights in John Finnis*). Persona y Derecho. 2008;**59**:138–144.

[6] Charter of the United Nations, signed in San Francisco, California on June 26, 1945. Available from: http://www.un-documents.net/ch-ppp.htm [Accessed: 2016-10-22].

[7] Orrego C. La 'gramática de los derechos' y el concepto de derechos humanos en John Finnis (*The 'Grammar of Rights' and the Concept of Human Rights in John Finnis*). Persona y Derecho. 2008;59:135–136.

[8] Gadamer H.G. Wahrheit und Methode (*Truth and Method*) (1960). 5th ed. Tübingen: Mohr; 1986. p. 387.

[9] Muñoz Conde F. La esterilización de los asociales en el nacionalsocialismo (*The Sterilization of Asocials in National Socialism*). In: López Barja de Quiroga J., Zugaldía Espinar J.M., editors. Dogmática y ley penal. Libro homenaje a Enrique Bacigalupo. Vol. 1. Barcelona: Marcial Pons and Instituto Universitario de Investigación Ortega y Gasset; 2004. p. 594, 607.

[10] Stolleis M. Recht mit Unrecht. Studien zur Rechtsgeschichte des Nationalsozialismus (*Law with Injustice. Studies on the Legal History of National Socialism*). Frankfurt am Main: Suhrkamp; 1994. p. 274.

[11] Baratta A. Rechtspositivismus und Gesetzespositivismus. Gedanken zu einer 'naturrechtlichen' Apologie des Rechtspositvismus (*Statutory Positivism and Legal Positivism. Thoughts on a 'Natural Law' Apology of Legal Positivism*). Archiv für Rechts- und Sozialphilosophie. 1968;**54**(3):327

[12] Neumann U. Rechtsphilosophie in Deutschland seit 1945 (*Philosophy of Law in Germany since 1945*). In: Simon D., editor. Rechtswissenschaft in der Bonner Republik. Studien zur Wissenschaftsgeschichte der Jurisprudenz. Frankfurt am Main: Suhrkamp; 1994. p. 147.

[13] Bessel R. Germany 1945. From War to Peace (2009). New York: Hape Perennial; 2010. p. 9.

[14] Weindling P.J. Nazi Medicine and the Nuremberg Trials. From Medical War Crimes to Informed Consent. New York: Palgrave Macmillan; 2004. p. 101.

[15] Mitscherlich A., Mielke F., editors. Medizin ohne Menschlichkeit. Dokumente des Nürnberger Ärzteprozesses (*Medicine Without Humanity. Documents of the Doctor's Trial in Nuremberg*) (1949). Frankfurt am Main, Hamburg: Fischer Bücherei; 1960. p. 280–281. Some names were missing in that process like Max Crinis or Carl Schneider, absent because they committed suicide in 1945, and Josef Rudolf Mengele himself, who disappeared without a trace, just to name a few.

[16] Weindling P.J. Nazi Medicine and the Nuremberg Trials. From Medical War Crimes to Informed Consent. New York: Palgrave Macmillan; 2004. p. 106.

[17] Weindling P.J. Nazi Medicine and the Nuremberg Trials. From Medical War Crimes to Informed Consent. New York: Palgrave Macmillan; 2004. p. 251, 254, 256.

[18] Zolo D. La giustizia dei vincitori. Do Norimberga a Baghdad (*Victors' Justice: From Nuremberg to Baghdad*). Roma: Laterza; 2006. p. 142.

[19] Hassenfeld I.N. Doctor-Patient Relations in Nazi Germany and the Fate of Psychiatric patients. Psychiatric Quarterly. 2002;**73**(3):184.

[20] Some could even continue their business quite normally like W. Bayer, H. Hefelman, W. Heyde, J. Mengele, O. von Verscheuer or H. J. Sewering, despite his nazi past. Proof of this was the obituary, published in the journal Deutsches Ärzteblatt by Hoppe J.D., Vilmar C., in which it opts for silencing the participation of the doctor in Hitler's program of 'euthanasia'. Hans Joachim Sewering†. Gestalter im Dienst der Ärzteschaft (*Designer at the Service of Medical Profession*). Deutsches Ärzteblatt. 2010;**107**(28–29):1409.

The reaction to this obituary did not take long: Nachruf: Kein Hinweis auf die Rolle im Nationalsozialismus (*Obituary: No Reference to the Role in National Socialism*) was published in the same journal signed by Hohendorf G., Fangerau H., Wahrig B., and other 78 professionals. They criticized the incomplete nature of the obituary, which might mean a step backward when it comes to present the reality of the facts, Deutsches Ärzteblatt. 2010;107(31–32): 1520.

[21] Klemperer V. LTI. Notizbuch eines Philologen (1949). It is quoted by the English edition The Language of the Third Reich. London, New York: Bloomsbury Academic; 2013. p. 2.

[22] The Nuremberg Code. Date: August 20, 1947. Available from: https://history.nih.gov/research/downloads/nuremberg.pdf [Accessed: 2016-08-19].

[23] Already at the beginning of the 1930s it was issued a document in a similar vein titled Forschungsrichtlinien des Reichsinnenministeriums *(Research Guidelines of the Reich Interior Ministry)*. Richtlinien für neuartige Heilbehandlung und für die Vornahme wissenschaftlicher Versuche am Menschen (*Guidelines on Innovative Therapy and Scientific Experimentation Involving Human Subjects)*. Deutsche Medizinischen Wochenschrift. 1931;57:509.

[24] Kaufmann A. Zur ethischen und strafrechtlichen Beurteilung der sogennanten Früheuthanasie (*On the Ethical and Criminal Assessment of the So-called Early Euthanasia*) Juristenzeitung. 1982;**37**:481.

[25] Lombardo P.A. Three Generations, No Imbeciles: Eugenics, the Supreme Court, and Buck v. Bell. Baltimore: The Johns Hopkins University Press; 2008. p. 239.

[26] Lombardo P.A., Dorr G.M. Eugenics, Medical Education, and the Public Health Service: Another Perspective on the Tuskegee Syphilis Experiment. Bulletin of the History of Medicine. 2006;**80**(2):292.

[27] Beauchamp T.L., Childress J.F. Principles of Biomedical Ethics. New York: Oxford University Press; 1979. 314 p.

[28] Adorno T.W. Negative Dialektik (*Negative Dialectics)*. Frankfurt am Main: Suhrkamp; 1966. p. 352.

[29] Mate R. Tratado de la injusticia (*Treaty of Injustice)*. Barcelona: Anthropos; 2011. p. 39.

[30] Ricouer P. La mémoire, l'histoire, l'oubli (*Memory, History, Forgetting)*. Paris: Seuil; 2000. p. 37.

[31] Mate R. Tratado de la injusticia (*Treaty of Injustice)*. Barcelona: Anthropos; 2011. p. 173.

[32] Ricoeur P. La mémoire, l'histoire, l'oubli (*Memory, History, Forgetting)*. Paris: Seuil; 2000. p. 100.

[33] On the rupture between memory and history. Ricoeur P. La mémoire, l'histoire, l'oubli (*Memory, History, Forgetting)*. Paris: Seuil; 2000. p. 523–524.

[34] Bueno G. Sobre el concepto de 'memoria histórica común' (*On the concept of 'common historical memory'*). El Catolepas. 2003;**11**:2.

[35] Mate R. Tratado de la Injusticia (*Treaty of Injustice*). Barcelona: Anthropos; 2011. p. 180.

[36] Mate R. Por los campos de exterminio (*For the Death Camps*). Barcelona: Anthropos; 2003. p. 172.

[37] Ricoeur P. La mémoire, l'histoire, l'oubli (*Memory, History, Forgetting)*. Paris: Seuil; 2000. p. 73–77, 453, 538, 553.

[38] Arendt H. Eichmann in Jerusalem. A Report on the Banality of Evil. New York: Viking Press; 1963.

[39] Kaufmann A. Über die Tapferkeit des Herzens (*About the Bravery of the Heart*). Archiv für Rechts- und Sozialphilosophie. 1991;**77**:16.

[40] Klemperer V. LTI. Notizbuch eines Philologen (1949). It is quoted by the English edition The Language of the Third Reich. London, New York: Bloomsbury Academic; 2013. p. 16.

[41] Goebbels J. Der Nazi-Sozi. Fragen und Antworten für den Nationalsozialisten (*The 'Nazi-Sozi': Questions & Answers for National Socialists*). Elberfeld: Nationalsozialistische Briefe; 1927. pp. (no page).

[42] Mate R. Por los campos de exterminio (*For the Death Camps)*. Barcelona: Anthropos; 2003. p. 132.

[43] Bessel R. Germany 1945. From War to Peace (2009). New York: Hape Perennial; 2010. p. 278.

[44] Weindling P.J. Nazi Medicine and the Nuremberg Trials. From Medical War Crimes to Informed Consent. New York: Palgrave Macmillan; 2004. p. 133–134. More in detail about the victors' justice, p. 161–165.

[45] Wiesel E. Without Conscience. The New England Journal of Medicine. 2005;**352**:1513.

[46] Adorno T.W. Erziehung nach Auschwitz (*Education after Auschwitz*) (1966). In: Tiedemann R, editor. Gesammelte Schriften. Vol. 10, part 2. Frankfurt am Main: Suhrkamp; 1966. p. 674, 683, 690.

[47] Arendt H. Eichmann in Jerusalem. A Report on the Banality of Evil. New York: Viking Press; 1963. p. 253.

[48] Habermas J. Geschichtsbewusstsein und posttraditionale Identität (1987). It is quoted by the English edition Historical Consciousness and Post-Traditional Identity. In: Habermas J. The New Conservatism: Cultural Criticism and the Historian's Debate. Cambridge: MIT Press; 1989. p. 251–252.

[49] Mate R. Por los campos de exterminio (*For the Death Camps)*. Barcelona: Anthropos; 2003. p. 66.

[50] Giubilini A., Minerva F. After-birth Abortion: Why Should the Baby Live?. Journal of Medical Ethics. 2013;**39**(5):261–263

Chapter 7

Ethical Issues in Organ Procurement and Transplantation

Richard J. Howard and Danielle L. Cornell

Additional information is available at the end of the chapter

http://dx.doi.org/10.5772/64922

Abstract

The Ciba Foundation held the first international, interdisciplinary conference on ethical and legal issues in transplantation in March 1966. Many of the ethical issues discussed at that conference remain with us today. Organ procurement and transplantation have forced the medical community and society at large to ask such fundamental questions as when are we dead, how can death be declared so that any life-support measures can be discontinued? Is it ethical to remove an organ or part of an organ from a living person? Since there is such a shortage of organ and people on transplant waiting lists die for lack of an organ, what types of incentives, if any, can be used to increase the organ supply? Transplant centers face additional ethical issues. How can a limited supply of organs be fairly allocated to a large number of patients on the waiting list? Are the methods of putting patients on the waiting list appropriate? Transplant centers are regulated by a variety of governmental organizations. These organizations may have performance criteria. Do these performance criteria lead transplant centers to modify which organs they will accept or which patients they will list? As long as a shortage of organs remains, these ethical issues are likely to persist.

Keywords: organ procurement, transplantation, organ allocation, brain death, organ donors, donor registries, international organ trade, transplant waiting list, transplant regulation

1. Introduction

The first successful organ transplant occurred in 1954 when Dr. Joseph Murray and his team in Boston transplanted a kidney between identical twin brothers. This was not the first kidney transplant reported. Dr. Yu Yu Voronoy reported, in 1936, a kidney transplant using a deceased donor [1]. That and several kidney transplants by Dr. David Hume in the early 1950s, also in Boston, were not successful. Once the success of transplantation was demonstrated and the development of immunosuppressive drugs permitted survival of organs from non-identical twins, transplantation rapidly developed. It was not long before the number of people needing organ transplants outpaced the supply of organs. The disparity between need and supply—currently in the United States, there are more than 120,000 people awaiting an organ transplant yet in 2015 there were only just over 30,000 transplants performed—has given rise to the issues that generate the most ethical discussions in organ procurement and transplantation: (1) How can we increase the number of both living and deceased organ donors and (2) What is the best way to allocate (ration) the scarce organs that are available? Twenty people on the waiting list die every day. Many others are removed from the waiting list because they become too sick while awaiting an organ. And things may even be worse. Because of the organ shortage and regulatory oversight of outcomes, transplant centers feel forced to list only the best candidates. Probably, many others could benefit from an organ transplant. Regulatory oversight of transplantation and organ donation with its performance requirements may have also contributed to having fewer organs for transplantation and fewer transplants [2].

The situation is similar in Europe and many other countries with a great disparity between the number of organ donors and the transplant waiting list. In many other countries, there is no or only minimal access to organ transplantation. These disparities between the number of donors and the number of patients on the transplant waiting list gives rise to several potential ethical issues.

2. Ethical issues in organ donation

2.1. Declaration of death

The first kidney donor was an identical twin. Since that time both living and deceased donor organs have been used. Initially, living organ donation was limited to blood relatives, because it was believed that there would be less likelihood or rejection. Since that time, immunosuppression has become so effective that currently virtually any healthy adult can donate. Kidneys are the most common organ from living donors, but livers, lung, intestine, pancreas, and even one heart (from an individual with a healthy heart who had it removed as part of a combined heart–lung transplant) have been performed.

Most organs, however, come from deceased donors. From the beginning, transplant programs have followed the dead donor rule (DDR): organs should only be recovered from donors who have died and nothing should be done by transplant or organ recovery programs that would hasten the death of a donor. There are two ways someone can be pronounced dead: by neurologic criteria and by circulatory criteria. Donation after neurologic determination of death (DNDD), also called brain death, occurs when the entire brain including the brain stem has no detectable function. Some physicians also complement the clinical examination with a radiologic test that shows no blood flow to the brain. Donation after circulatory determination of death (DCDD), also referred to donation after cardiac death, occurs when there is no detectable blood flow in a patient. There is usually a waiting time of 2–5 min to ensure that

http://dx.doi.org/10.5772/64922

blood flow will not spontaneously reoccur, but some hospitals may use a shorter or longer waiting period. Both measures for determining death have raised concern.

The first kidney transplants used organs from donors who died from circulatory death [1]. The concept of neurologic or brain death did not come about until the late 1950s. It arose with the development of ventilators and intensive care units. There were patients who demonstrated no clinical brain function but who when maintained with the aid of a ventilator continued to make urine, have a normal blood pressure, and exhibit many normal physiologic findings. Yet, these individuals never recovered. That caused physicians, ethicists, philosophers, religious leaders, and government officials to ask what it really means to be alive or dead. Ultimately, the concept of neurologic death came about, albeit with much controversy. The concept of neurologic death changed the criteria for determining death that had been used for thousands of years. Many physicians and ethicists initially believed it was not ethical to consider neurologic death. And, the new concept seemed to violate traditional religious concepts of determining death.

Some individuals (and some cultures and religions) feel it is wrong to "desecrate" the body of a dead person by using the body parts for transplantation. But this is a minority view. The three Abrahamic religions—Christianity, Islam, and Judaism—while believing that a dead body should not be desecrated and should be treated with great respect, all feel that preserving life through organ transplantation is a higher good. Therefore, recovery of organs for transplantation is not only permitted, it is to be commended [3].

Virtually, all hospitals are required to have brain death policies. But brain death is not likely to be pronounced in hospitals without intensive care units or that do not have the ability to ventilate patients such as rehabilitation hospitals, psychiatric hospitals, and hospitals without intensive care units. The criteria used to diagnose brain death may vary from state to state. Some states require only one physician to pronounce brain death, while others require two physicians to make the diagnosis before death is declared. The definition of brain death is frequently according to generally accepted medical criteria. One set of criteria has been defined by the American Academy of Neurology [4].

Ultimately, death determination by neurologic criteria became accepted by most individuals, including religious experts. The use of organs from DCDD donors was greatly reduced and, in many places, was eliminated altogether. But growing shortage of organs sparked renewed interest in DCDD. In the United States, several organ procurement organizations (OPOs) began to recover organs from these donors, because transplant centers were willing to use them. The same changes occurred in many other countries as well. But now, the medical and lay communities had grown so used to organs recovered from DNDD donors that some individuals objected to using organs from DCDD. They asked whether these donors were really dead. The United Network for Organ Sharing has published a set of guidelines for recovery of organs from DCDD donors: Model Elements for Controlled DCD Recovery Protocols (https://www.unos.org/about/governance/).

The controversy about the ethics of DCDD and DNDD continues to this day. Some people question whether these donors are truly dead. Rady et al. [5] have reviewed the arguments against the validity of neurologic and circulatory death. Objections to neurologic death include that neurologically dead patients maintain many functions that are coordinated by the brain. There is often preservation of hypothalamic–pituitary functions. A non-flat-line electroencephalogram (EEG) can frequently be obtained. Clinical tests to confirm irreversible cessation of whole brain and brain stem functions are not completely reliable. But brain-death determination requires that the function of the brain and brain stem should be permanently absent. It does not require that every cell in the brain be dead. The EEG is unreliable as a determinate of cerebral activity, since it is sensitive to surrounding electrical activity in the environment.

Objections to using circulatory death determination include the possibility of spontaneous autoresuscitation or the presence of electrical activity of the heart after 10–15 min of circulatory arrest, preservation of cerebral activity on EEG after three minutes of circulatory arrest and that hearts recovered after circulatory arrest can be successfully transplanted. There are other objections as well. Heparin, an anticoagulant, is usually given to the donor just before death occurs to prevent clotting in the organs. Heparin could theoretically hasten death if it induces bleeding into the brain. But since it is given only in the agonal stages of life immediately before death, it is unlikely to hasten death [6]. If the donor does not die within 60 min after withdrawal of ventilator support, recovery surgeons usually do not recover the organs, because the hypotension and lack of oxygen in the blood may have damaged the organs. But with new preservation techniques, surgeons are waiting up to 120 min and even longer in some cases. The waiting time after cessation of circulation may not permit sufficient time for the brain to die.

Because of difficulties some ethicists had with issues involved in assuring those declared dead by either neurologic or circulatory criteria are really dead, they feel that the dead donor rule and current methods of declaring death were accompanied by illusions and myths and that organ recovery may hasten the death of the donor [7–9]. They propose that the DDR should be abandoned and that a different recovery process should be initiated in patients with severe, unrecoverable brain injuries. Patients with devastating brain injuries, who would otherwise be declared brain dead or withdrawn from life support, should have their organs recovered under general anesthesia as part of the process of withdrawal of life support [5]. In their view, such recovery would avoid the fiction of declaring death by current methods and permit the recovery of more organs for transplantation.

2.2. Living donors

Living donors are those individuals who are alive at the beginning of organ donation and are expected to be alive at the end. They donate one organ or, in some cases, a portion of one organ. There were approximately 6000 living donors and 9000 deceased donors in the United States in 2015. Removal of an organ from a living donor is the only operation that is done specifically to help another person. In fact, living organ donors are chosen because they are healthy. So living organ donation would seem to violate one of the tenants of medicine: *primer non nocere,* first do no harm. Although they cannot be physically better off from the donation, they can receive psychological benefit from having helped a loved one or another person they may not even know.

http://dx.doi.org/10.5772/64922

But living organ donation has the potential to do great harm. First, there is the risk of death from the operation itself. The risk do dying during donor partial hepatectomy is 0.2%, and the risk of death during nephrectomy is 0.03% [10, 11]. The mortality risk of donor pancreatectomy or lung donation has not been determined. Even if the donor survives the operation, long-term health outcomes may still be at risk. Some liver donors have had to have liver transplants themselves because of injury to the bile duct. Most of the long-term outcome results have been recorded in kidney donors. Some reports from the United States have reported no adverse effects on longevity, hypertension renal function, or quality of life with follow-up as long as 40 years after donation [12]. Other studies, however, report a higher long-term mortality and risk of renal failure in kidney donors followed for as long as 44 years [13]. In the United States, data regarding living kidney and liver donor complications are available in reports of the Scientific Registry of Transplant Recipients [14].

Because of the real and potential risks to living organ donors, it is critical that these donors be carefully selected so as to minimize the short-term and long-term risks. In fact, Moore [15] questioned many years ago whether it was ever morally right and ethically acceptable to injure one person to help another. A statement by the Live Organ Donor Consensus Group did provide some guidelines for the evaluation and education of potential living donors [16]. Every transplant center would like to use only people who are in a state of perfect health and have normally functioning kidneys. But in an evaluation of the Organ Procurement and Transplant Network (OPTN) database, Davis and Cooper [17] found that 19.5% of donors were obese, 2% had a history of hypertension, 3.5% had proteinuria, and 12.2% had no health insurance. So one concern is whether it is ethically justifiable to use anyone as a living donor who is in less than perfect health, and, if so, how far can a donor deviate from less than perfect health?

Transplant centers differ greatly in their living donor acceptance criteria. And, the transplant centers that do accept (as most do) donors with less than perfect health can have different criteria. What degree of renal dysfunction or hypertension would disqualify a potential candidate? And on what basis does the center make its decision other than personal feeling of the selection team. Grams et al. [18] have recently created a model that can help to predict the risk of eventual renal failure in potential kidney donors. But there will always be some uncertainty in donor selection. Steiner [19] has pointed out that a "safe" kidney is a delusion. Centers should not ignore risk quantification, however, imperfect and only do what they are "comfortable" doing, because they must have defensible reasons for doing what they do. It is unethical to allow or to deny transplants without good reason.

It is important that potential living donors be fully evaluated and psychologically screened. It is possible that family members may have pressured someone to donate an organ. If the transplant team learns that the person does not really want to donate, many transplant centers will provide a medical excuse that eliminates him as a donor and does not alienate him from the rest of the family. Yet providing such an "excuse" involves intentional dishonesty by the transplant center. Is such behavior ever ethical? Even without pressure from family members, Testa [20] has raised the issue of whether donation can ever be completely voluntary because of the emotional relationship to a family member who needs an organ, something he has called "pressured consent."

Transplant centers are understandably wary with donors who are not relatives of the recipient or even have no relationship and who are willing to donate to whomever needs a kidney. The United States, Canada, and some European countries have recognized these non-related donors or altruistic, although their evaluations mat receive special psychological scrutiny. Guidelines for non-directed donation have been developed in the United States by the United Network for Organ Sharing and the American Society of Transplant Surgeons [21].

Because of the unique nature of organ donation, it is important that the surgical team that cares for the donor be completely separate from the team that cares for the recipient. A consensus document advocated donor counselors who are independent of the transplant center to prevent the self-interest of the transplant center from influencing their judgment. Independent counselors or donor advocates could prevent the donation if they detected any unacceptable psychological or medical issues.

However, it is accomplished, donor autonomy must be preserved, and the donor must be fully informed of potential short-term and long-term risks. And yet, transplant centers may vary in how they present risks. They can cite papers showing that kidney donors live longer or shorter than non-donors or that there is no difference. Many donors have made up their minds to donate before their first visit to the transplant center and are not easily dissuaded by any discussion about risks. Nevertheless, it is incumbent on every transplant center to present an honest and factual assessment of donor risks to the best of its ability.

2.3. Financial incentives

One of the most contentious issues is whether incentives—financial or otherwise—should be used to increase organ donation. In the United States, the National Organ Transplant Act (NOTA) has made offering organs for "valuable consideration" illegal. It is illegal to sell organs in every other country as well except for Iran. Saudi Arabia had an experimental trial of financial incentives in Riyadh that showed that it could substantially increase organ donation [22]. Yet, the public in the United States are in favor of some sort of reimbursement [23]. Until recently, living donors endured expenses such as travel, lodging, meals, and lost wages. Recently, in the United States, the National Living Donor Assistance Center (www.livingdonorassistance.org) began making help available help for travel and lodging but not for lost wages.

Those who argue against incentives for organ donation point out that the altruistic system currently in place has served transplantation well. They maintain that having only altruistic donors—whether related to the donor or not—has eliminated any sense of coercion. Opponents of incentives also emphasize the potential risk to donors and the impact incentives might have on society's moral perspective. They cite harms such as coercion, exploitation, undermining dignity, repugnance, and commodification [24]. They talk, however, about unregulated markets. Placing a price on body parts would lead to commodification of the human body, something that runs counter to religious teaching and good ethical practice. Other ethically appropriate methods should be used to increase donation. Furthermore, they argue, offering payment for organs would offend the sensibilities of many people and might result in fewer organs donated.

Those against financial incentives for living donors also claim that it is unethical for medicine to harm paid kidney donors for the benefit of others and that removing an organ, usually a kidney, leaves the person in poorer health. If that is so, of course, then removal of an organ for transplantation from any living donor would also be unethical. In some of these discussions, the international organ trade becomes conflated with financial incentives for living donors within a country.

Those in favor of incentives point out that the current system is not functioning adequately as attested by the long wait for an organ and an increasing number of deaths on the waiting list. Matas and Hays [25] favor regulated markets and maintain that a government regulated system would prevent the concerns voiced by opponents. Advocates of incentives claim those who are against having at least a trial of incentives are willing to see people on the waiting list die to preserve their own moral purity. Matas and Schnitzler [26] showed that living unrelated kidney donation would save more than $90,000 (2002 US$) and 3.5 quality-adjusted life years. They calculated that offering kidney "vendors" $90,000 would be the break-even point for society (because of the high costs of maintenance hemodialysis) and would greatly help alleviate the waiting list. They are in favor of a regulated system of payment; there would be no payment outside the regulated system. Matas and Hays [25] argue that with organ donors remaining steady for the last 10 years and that during those 10 years more than 60,000 candidates have been removed from the waiting list because of death or because they became too ill. They maintain that those in favor and those opposed to incentives should stop talking past each other and carry out a well-designed clinical trial of incentives.

Those in favor of financial incentives add that payment for living kidney donors has eliminated the waiting list in Iran, the one country in the world that permits payment for kidney donation (which may not be considered an actual donation). Iran has had a regulated system of payment for living kidney donors for several years [27, 28]. The government provided payment of a fixed sum ($3000–$6000) to the donor. In addition, he was exempted from military service and received 1 year of health insurance [28]. Recently potential donors and recipients have been allowed to bargain directly and to agree on an amount for the kidney. This system has eliminated the kidney waiting list. Indeed, the only waiting list is for those who want to be kidney donors. But kidney donors in Iran tend to be the poorest individuals from the poorest part of the country. And many of them have had complications and poor health outcomes. Many, however, look on the Iranian model as exploiting the poor to benefit the rich.

Financial and other incentives have also been proposed as a way to increase the number of deceased donors. Considerations such as providing payment to the donor's family, funeral expenses, donation to a charity, reduction in taxes, provision of health insurance, or preference on the waiting list should they ever need a transplant are some that have been suggested. All such incentives are currently illegal everywhere. But recently, the American Society of Transplantation and the American Society of Transplant Surgeons held a workshop on increasing organ donation [29]. They noted that NOTA was recently changed to allow certain expenses of living donors to be covered. While they believed NOTA did not permit direct payments for deceased organ donation, it could be interpreted to permit certain expenses of the donor or family to be covered. In a similar vein, a panel of ethicists, organ procurement organization executives, physicians, and surgeons reported in 2002 on financial incentives for deceased organ donation [30]. The panel addressed whether an ethically acceptable pilot trial of financial incentives could be proposed for the family to consent to the donation of organs from a deceased relative. While the panel was unanimously opposed to the exchange of money, either directly or in the form of a tax incentive, for donation, they believed that it would be ethical to provide funeral expenses or a charitable contribution that conveyed the appreciation of society to the family for the donation. And since the panel believed that direct financial incentives are unethical, there could be no ethical trial of direct financial incentives. However, with several restrictions there could be a pilot trial of funeral expenses or a charitable contribution.

There are many arguments against financial incentives. There are also many arguments in favor of incentives. Many of the arguments against financial incentives have been summarized by Arnold et al. [30]. Generally, the arguments against financial incentives claim that they would violate altruism in organ donation and would commercialize organ donation in an unacceptable way by commodifying donor organs. Such commodification "uses the human body as a means rather than as an end in itself and brings an unacceptable commerce to the value of human life" [30]. They point out there is no evidence that financial compensation will increase organ donation (of course, except for Iran, it has not been tried).

Paying for organ donation would exploit the poor who would be most likely to accept financial incentives, and the organs would likely go to wealthier patients. But as some have pointed out, it is the poor who disproportionately are increased on the waiting list but who receive relatively fewer kidneys than wealthier waitlist candidates, and the poor die at a higher rate on the waiting list. Perhaps financial incentives, even if the poor are more likely to take advantage of them, will more likely help poor patients on the waiting list. An incentive system could increase the likelihood of transplanting organs from donors with diseases that might be transmitted to the recipient if the donor or family withheld important medical information in order to receive a financial incentive. Incentives could also influence the family to withdraw care prematurely. It could blur the line between withdrawing life support and donating organs [30]. It would also be difficult to standardize a system of payment and might introduce bargaining between donor families and recovery organizations, especially for organs of better quality and would place the transplant community on a pathway to paying for organs from live donors [30]. Those in favor of financial incentives or a "market" that would allow organ sales often offer an economic analysis of cost-effectiveness, but that still does not make the buying and selling of organs ethically acceptable. Furthermore, organ sales would likely not add additional organs for transplantation. Rather, sales or financial incentives would result in less altruistic donation. Organ sales would have an adverse effect on society and medical professionalism [30]. It would be difficult or impossible to set up a truly regulated market as many favoring incentives favor, because those in need of organs who have the means would likely make additional offers or go outside the system to make sure they got an organ.

There are also many cogent arguments in favor of financial incentives—if not totally changing the law, at least conducting a carefully regulated clinical trial [31–35]. The House of Delegates of the American Medical Association has come out in favor of a pilot study of financial

http://dx.doi.org/10.5772/64922

incentives for organ donation [36]. Those in favor of incentives or a market in organ sales point out that with current methods of altruistic organ donation, the number of donors is woefully inadequate to meet the need of those on organ waiting lists. The lack of organ donors becomes worse every year in relation to the wait list. And the waiting list may only represent a portion of those who could potentially benefit from an organ transplant. Transplant centers know there is a great shortage of organs and may only list their best patients. If many more organs became available, they might list additional patients who could also benefit from a transplant. The current altruistic system of obtaining organs has failed to supply enough to save the lives of thousands who could benefit. Even some of the most vocal opponents of financial incentives have favored a small reimbursement ($300), which is meant to express appreciation for the donation and not to provide payment for it [37]. But this "appreciation" is still a payment of money to the family of an organ donor and would not occur without the donation.

If an important part of a government's job is to preserve the lives of its citizens, and if financial incentives would increase the number of organs available, then government is not fulfilling its obligation. But by denying financial incentives because they would exploit the poor, harm the donor and commodify the human body, those who are against incentives claim to know what is best for potential donors and donor families. They can be accused of denying autonomy to the poor. We ask the poor to do many jobs that we ourselves would not want to do. And they may be anxious to do them, because they may not be qualified for more skilled, higher paying jobs. Furthermore, society does allow the sale of eggs, sperm, plasma, and the temporary sale (rental) of uteri for surrogate births. But those who are against incentives do not claim that all of these should be outlawed.

The issue of financial incentives either for living or deceased donors is not likely to be settled anytime soon. It could be that two ethical principles seem to clash. On the one hand, incentives might be bad for the reasons those opposed give. Yet the other ethical principle of valuing human life and attempting to preserve it by making more donors available with financial incentives is also a good argument.

2.4. Donor registries and authorization

Given that the current methods for determining someone is really dead are valid and that we are not breaching any ethical issues by recovering organs from them (despite that some continue to disagree—see above), what methods besides financial incentives can be used to increase the number of donors to help alleviate the great shortage? In 2015 there were 9080 deceased donors in the United States [38]. Yet, according to a study by Klassen et al. [39], there are more than 37,000 potential donors. No doubt, similar shortfalls in actual versus potential donors exists everywhere in the world.

In the United States as well as many other places in the world, organ donation has long relied on opting in; the donor while alive or the family has to grant permission for organs to be recovered after death. Some have argued that we all have a duty or moral obligation to our fellow man (or woman) to donate our own organs after death or those of a loved one who is a potential candidate for donation [40–43]. Menzel [43] maintains "contributing cadaver organs is not a matter of charitable goodness but instead normally an instance of the moral duty of easy rescue." The main argument has been that donation (some do not even like the word "donation" because it implies that the person has a choice and is not obligated) after death does great good, causes no discomfort to the organ donor, and does not cost the donor and the family anything [40].

With opting in, the transplant community has relied on donor and family altruism to grant permission for organ recovery. Public education and pleas to stress the need for organs have been used to increase organ donation. There are many routes to educating the public about the need including newspaper and magazine articles, television and radio announcements, billboards, posters and cards in public venues, distribution of donor cards, teaching materials for schools, talks by health personnel or donor families or living organ donors, and appearances at health fairs [44]. In recent years, the internet has also been used to promote organ donation through social media and a variety of web sites. In the United States, Donate Life Month occurs in April. Many public and private organizations promote organ donation and educate the public about the need for organs. Education of key hospital is also needed so the OPO is called every time there is a potential donor.

People can sign to be on a donor registry at a driver's license facility or through the internet. Information is provided at the driver's license facility and on registry sites on the internet that are inform the public about what agreeing to be an organ donor entails. Registration only covers brain dead donors. For donation after circulatory death, the family must still give its agreement.

In recent years, donor registries have registered individuals during life who have indicated their willingness to be organ donors when they die. And the transplant community has gone to great effort to increase the number of people who sign up for donor registries, either when renewing their driver licenses, signing organ donor cards, or registering on the internet. In many states, signing organ donor cards or being on state registries enables organs to be recovered after death even if the family objects. Israel gives individuals who have agreed to be an organ donor if they die preference on the waiting list if they should ever need a kidney transplant [45].

Because large organ shortages continue with opting in systems, some countries have adopted other organ recovery strategies and still others have been proposed. These include opting out, mandated choice, and financial incentives.

The United States and most other countries have long had an opting in form of consent whereby the donor while alive or the family of a deceased potential donor must give consent for organ donation before organs can be recovered. But many countries have adopted an opting out (also commonly called presumed consent) strategy to increase organ donation. Veatch and Pitt [46] have pointed out that none of the laws in the countries that permit organ recovery without explicit consent of the donor while alive or family actually mentions the words presumed consent. Among countries with versions of opting out laws are Austria, Belgium, Finland, Italy, Norway, Spain, and Switzerland. Cyprus, Hungary, Singapore, and Syria also have laws authorizing organ recovery without claiming to have presumed consent [46]. In these countries, organ recovery can occur following death unless the family specifically objects or the

http://dx.doi.org/10.5772/64922

donor while alive has indicated while alive that he does not want recovery to occur, but organ recovery agencies do not have to ask permission. While presumed consent still permits families to object to donation and to prevent it from occurring, the onus is on them to take the initial steps. Presumed consent has led to an increase in organ recovery in many countries that have adopted it. Spain is the most successful country in the world in organ recovery. In addition to presumed consent, Spanish hospitals have physicians who are responsible for identifying organ donors and promoting donation in the intensive care units [47].

There are ethical and practical issues in adopting opting out or presumed consent. It is counter to current practice in the United States and requires new legislation. Some claim it would be more humane than opting in, because organ procurement coordinators would not have to discuss organ donation at a difficult time for the family. But it may overlook the family's knowledge of the individual's preferences and may increase distrust of the medical community with concern that death may be declared prematurely [44]. Presumed consent would also cloud who has control over a deceased person's body (there is no ownership rights to a dead body). Veatch and Pitt [46] prefer the term "routine salvaging" to presumed consent as being more honest, because that term refers to a policy that is not grounded in presumption, but rather in a belief that society has a right to recover organs without individual consent. In other words, Can the state maintain that one of its important functions is to preserve the lives of its citizens and therefore can salvage (some would say confiscate) dead bodies for the purpose of organ recovery?

Mandated choice would require every adult to decide and record whether they wish to be an organ donor when they die. No country currently has a policy of mandated choice. It would eliminate the need to obtain consent. It would relieve the family of what might be an agonizing decision and prevent family disagreement. Because the person made a decision about their body before death, many of the objections of presumed consent would be obviated. Mandated choice has been criticized as being insensitive to families and forcing individuals to confront their own death. The American Medical Association Council on Ethical and Judicial Affairs feels mandated choice is not coercive because individuals are free to say no [48]. And by requiring individuals to decide on donation, it promotes autonomy.

2.5. The international organ trade

Many people, usually from First World countries and wealthy individuals from other countries among which are European nations, the United States, Japan, and the Gulf States, travel to Second and Third World countries where they can purchase kidneys and receive transplants. Also called transplant tourism, this travel is due to the insufficient number of organs in their home countries. In 2007, the World Health Organization estimated that of the 60,000 kidney transplants performed annually around the globe, 5–10% were due to the international trade in organs [49]. In 2011, the Institute, Global Financial Integrity, ranked the international organ trade in the top 10 of the world's most profitable crimes, with an estimated profit of $614–$1200 million per year [49]. The European Union has funded the formerly Human Organ Trafficking for Transplantation (HOTT) project, which addresses "trafficking in human beings for the purpose of organ removal" [49]. With the primary concern being exploitation of the poor, virtually every country has made the international organ trade illegal and numerous international organizations have written position papers condemning it including the Transplantation Society, the Council of Europe, the World Medical Association, the Bellagio Task Force, and the Declaration of Istanbul (by the Transplantation Society and International Society of Nephrology, the World Health Organization, and the International Congress on Transplantation in developing countries) [50–52]. But transplant tourism continues to occur.

In addition to the ethical arguments against paying for organs given above, the international organ trade is also condemned because of the types of people willing to sell organs and the negative effects kidney removal has on the organ sellers. It is not ethical to take advantage of poor organ sellers for the advantage of the rich. People willing to sell their kidneys are poor and live in the poorest parts of the poorest countries. Their desperate situations mean that the choice to sell a kidney is not free or autonomous [42]. The international organ trade results in transferring kidneys from the poor to the wealthy who can afford to pay for transplants and the travel to countries where they can receive a transplant. These desperate kidney sellers may not provide a complete medical history either from ignorance or from fear of being rejected as a donor, which can further lead to adverse effects on their own health or to transfer of disease to the kidney recipient. After the kidney removal, the sellers have numerous problems including hypertension, kidney insufficiency, infection, and other medical problems [52]. They have limited, if any, access to medical care should they need it. They frequently become unemployable, because they are usually unskilled and not able to sustain heavy agricultural or construction work. Furthermore, they may become social outcasts and are alienated from their families, excommunicated from their churches, and excluded from marriage.

But there are opposing viewpoints that argue in favor of allowing individuals in any country to sell their kidney to willing buyers. Bakdash [52] who himself grew up in poverty says that "*poverty itself* is a kind of coercion. None of the decisions any poor person makes is made on the basis of free will—instead, these decisions are all dependent on the person's dire financial situation." He points out that the desperately poor may have to choose between selling a kidney and letting their children starve. People may be willing to sell their kidneys because they want a chance at a better life.

All the condemnations by the numerous health and transplant organizations have not stopped the buying and selling of organs in Second and Third world countries. While those who oppose buying and selling of organs point out that poor organ sellers frequently get poor medical care, Bakdash [52] points out that if medical care for these individuals were taken out of the shadows and brought into the open through a regulated market, they would be able to avail themselves of better medical care. Radcliffe-Richards and colleagues [53] have pointed out that there is much greater opportunity for "exploitation and abuse when the supply of desperately wanted goods is made illegal."

Others maintain that prohibiting organ sales takes away the potential seller's autonomy and is paternalistic. They believe that the wealthy who write the rules and regulations find it convenient to tell the poor what is good for them and to deny them the opportunity to possibly improve their situation.

2.6. Ethical issues in transplantation and organ allocation

As long as there is a shortage of organs for transplantation, there will be a requirement that those organs be offered to patients on the waiting list in an ethical manner. The allocation system should be "fair." The Organ Procurement and Transplant Network/United Network for Organ Sharing (OPTN/UNOS) Ethics Committee adopted and updated in 2015 a white paper, "Ethical Principles in the Allocation of Human Organs" [54]. These principles provide a framework for regulations for the organ allocation policies.

The three ethical principles that govern organ allocations policies are as follows: "utility (doing good and avoiding harm), justice, and respect for persons" [54]. Utility "refers to the maximization of net benefit to the community (taking into account both the amount of benefit and harm and the probability of such benefit and harm)". Justice refers to the fair distribution of benefits. And respect for persons refers to telling the truth, keeping commitments, and, especially, respect for autonomy. The OPTN/UNOS Ethics Committee realized that recommendations are for policy in a pluralistic society in which individuals may hold conflicting, yet reasonable, positions on organ allocation.

The three ethical principles individually may lead to policies that conflict with each other. Therefore, the principles have to be balanced in order to achieve an equitable outcome. Utility should lead to maximizing the net benefit for the community, thus incorporating the ethics principles of beneficence (doing good) and non-maleficence (not doing harm). In maximizing utility factors such as patient survival, graft survival, quality of life, alternative treatments, and age can be taken into account [55]. Social aspects such as social worth, social status, occupation, race, and so forth should not be considered in formulating policy.

The OPTN/UNOS Ethics Committee uses justice to refer to "fairness in the pattern of distribution of the benefits and burdens of an organ procurement and allocation program" [54]. It does not mean treating all patients the same but does "require giving equal respect and concern to each patient". Factors to be considered in the application of justice include medical urgency, likelihood of finding a transplant in the future, wait list time, first versus repeat transplant, age, and geography. Autonomy requires treating people as ends in themselves, not only as means. But sometimes, respect for autonomy conflicts with other ethical principles, and sometimes, autonomy must be respected and other times, it must give way.

These ethical principles may lead to conflict when it comes to formulating actual policies. The OPTN/UNOS white paper also provides guidelines for resolution among principles. As a compromise when ethical principles conflict, an attempt should be made to formulate policies that give each of the conflicting principles equal weights rather than ranking them in some order.

Even though the white paper serves as an important set of guidelines, there have been strong disagreements about how to put them into practice. The OPTN has periodically changed the allocation scheme for every organ, always trying to achieve a new policy that would be more fair and better reflect the ethical principles. Yet as long as the organ shortage remains (which it will for the foreseeable future), any change in policy that increases organs to one group of patients must take away from another group. The dilemma of selecting ethical allocation schemes was shown in recent years in the United States with the proposed revision kidney allocation using projected survival after transplantation and kidney quality and in 2011 to use kidney quality and allocating 20% of kidneys to patients with the highest estimated post-transplant survival. The current system, based primarily on waiting time, is patient centered. The proposed system which recognizes the value society places on the life-extending potential of a scarce resource is resource centered. Proponents and opponents of this proposed allocation scheme use practical and ethical arguments to support their positions. For instance, one reason proponents are in favor is because it would lead to more life years in recipients and would direct more kidneys to younger patients who have not had the opportunity to live as long as older individuals on the waiting list, while opponents object because the new allocation scheme would result in age discrimination.

Other alterations to the allocation of kidneys to favor the young by giving them primary access to kidneys from younger donors that are regarded to be better quality than kidneys from older donors is disadvantages to older patients on the waiting list. Allocation schemes that favor multiorgan transplants over single organ transplants disadvantage patients who would otherwise have received the second organ. Directing kidneys preferentially to highly sensitized patients with high panel reactive antibody may be good for this set of difficult-to-transplant patients but directs kidneys away from patients with lower antibody levels. Directing organs to maximize patient survival may lead to one set of allocation schemes whereas allocating organs to the sickest (and thus preventing imminent deaths of other patients) could favor a different allocation scheme. All these allocation schemes by favoring one subset of patients may be good for society overall, but they necessarily discriminate against other groups of patients on the waiting list. Should patients who have had one transplant that has failed be given a second transplant when so many patients have not even had a single transplant?

As Chumfong and colleagues [55] said, "all allocation systems ought to achieve a version of distributive justice for the good of society. The fundamental issue at odds in the current and new allocation systems is what exactly the good of society is."

While the current discussion is mainly from the perspective of the ethical issues involved in kidney allocation in the United States, these same issues apply to virtually every other country and to other organs as well. As long as the shortage of organs continues, good, well-meaning people will disagree on precisely what form the best and most ethical policy is. This disagreement may stem from their unique situations in life and work, their backgrounds, culture, perhaps even their genetics. What is interesting, although subject for a different discussion, is just why people who agree on ethical principles may disagree vehemently on their actual application. Nevertheless, the allocation system in the United States is always a work in progress as UNOS constantly strives for more fairness for patients, better outcomes, and minimizing wasteful discarding of transplantable organs.

2.7. The waiting list

Currently, individual transplant centers determine which patients are placed on the waiting list. There may be guidelines, but there are few established criteria. The actions of individual transplant centers are important, because who is on the list affects who gets transplanted and therefore how organs are allocated. Thus, important ethical considerations can affect allocation before an organ even becomes available.

Both medical and non-medical criteria are used in deciding whether a potential candidate should be placed on the waiting list. Each organ may have its own set of criteria. For instance, to be placed on the waiting list for a kidney transplant, patients frequently have their renal function and bladder function evaluated. Some patients may also undergo cardiac evaluation, assessment of immune status against certain infectious agents, etc. Transplant centers frequently differ in how they use the results of medical testing in their decision to list patients. Patients may be turned down at one transplant center but accepted at another, leading some patients to "shop" for a transplant center willing to list them. Differences in medical criteria between transplant centers may not pose ethical issues; they may just represent honest disagreements between centers [56].

Transplant centers also evaluate patients using behavioral and other non-medical criteria. Virtually, everyone agrees that group characteristics such as religion, ethnicity, race, etc. should not enter into the decision of whether to place a patient on the waiting list. The UNOS Ethics Committee has recently addressed the non-medical considerations in assessment for transplant candidacy [54]. While age or co-morbidity should not arbitrarily be used as criteria for listing patients, life expectancy with a functioning graft using factors such as age, co-morbidities, and other factors can used if it is significantly shorter than the expected life span of the transplanted organ. But these decisions based on age and comorbidities should be made on an individual patient basis.

Transplant programs frequently refuse to list patients who exhibit some behavioral characteristics such as smoking, addiction, drug abuse, history of noncompliance with a medical regimen, or mental disability. Caplan [57] has written that such exclusion of categories of patients such as these increases doubt about the equitable allocation of organs. But transplant centers that do use these categories to exclude patients from listing justify their actions on the basis of being good stewards of precious organs. It would be a tragedy to transplant an organ and have it rejected because the patient did not take her antirejection medications because of behavior characteristics that were known before listing. Every transplant center has had experience with patients losing organs for these reasons. If this type of loss occurs, it means the recipient had an operation for no benefit (and likely would need a second operation to remove the organ or a second transplant) and another patient on the waiting list was denied access to that organ. Furthermore, the donor family who often are in contact with the recipient of their loved one's organs may experience a second loss if the transplanted organ is lost. This exclusion is especially true if a behavior contributed to organ failure such as alcohol consumption and liver failure or smoking and lung failure. These transplant centers require patients to quit using alcohol and smoking, although they may disagree how long abstinence is required before listing. And some centers permanently remove such patients from the waiting list should they relapse. Caplan [57] might think that is too harsh and would lead to doubt about the allocation of organs (since allocation begins with who is put on the waiting list). But most transplant centers believe such exclusions are good medical practice and the correct ethical decision.

Behaviors that may exclude patients from listing may not be permanent. The ethical issue for transplant centers is whether patients can recover from what they consider to be unacceptable behaviors and what criteria the patients must demonstrate to show they are now an acceptable candidate. An issue for the UNOS Ethics Committee is whether non-compliance behavior is serious, consistent, and documented in current or previous treatment [54].

2.8. Transplant volume and regulation

Other ethical issues for transplant centers arise at the juncture maintaining quality, increasing numbers of transplants, minimizing costs for the hospital, and satisfying regulators [58]. Transplant physicians and surgeons generally want to perform as many transplants as they can in order to serve their patients on the waiting list and to increase the status of the program. This may cause them to place patients on the waiting list who are far from ideal candidates. And because many transplant candidates have other health issues in addition to the primary organ failure, the issue becomes how far can one deviate from an ideal candidate and not pose too great a risk both for the patient and the transplant program. With too great a set of qualifications to be on the waiting list, the program may deny access for many qualified candidates. But too few qualifications also are unfair to the patient because of the excessive risk, it may impose from transplantation. There are no firm guidelines and programs differ in their qualifications for listing.

Similarly, transplant programs must accept organs that have a high certainty of function when transplanted. In order to increase the number of transplants, it is tempting for transplant centers to accept organs that are less than ideal. But how far can an organ depart from ideal and still be ethically acceptable to transplant? Here, again there are no firm guidelines and transplant centers differ in organ acceptance criteria. Patients are supposed to be informed if the organ is less than ideal, but most patients have limited ability to appreciate the many subtleties that go into the decision to accept an organ. But it is more likely that transplant centers turn down organs that can be transplanted. Many organs are discarded that are suitable for transplantation [59].

Listing patients for transplantation and accepting organs are influenced not only by the patient's need for a transplant. Also important, although not frequently mentioned, are other important influences that affect these decisions. Hospital administrators want transplant programs that are successful and do not operate at a loss. They want a high-quality program that performs many transplants. They may pressure transplant physicians and surgeons to minimize hospital stay to shorter than they think is best for the patient, use less expensive medications, and to transplant only well-insured patients, and not place higher risk patients on the waiting list because they usually have much higher costs. They may not express these feelings overtly, but the transplant physicians and surgeons usually get the message very clearly.

Regulators also play a role in creating ethical dilemmas. The Centers for Medicare and Medicaid Services (CMS) in the United States, which regulates and funds much of transplantation, in 2006, published the Conditions of Participation for Transplant Centers [60]. The

http://dx.doi.org/10.5772/64922

Conditions of Participation for Organ Procurement Organizations were published a year later [61]. The conditions of participation (COP) for transplant centers had performance criteria with expected outcomes. Transplant centers could be closed if their outcomes were below expected levels. This threat may have caused transplant centers to become more conservative in their acceptance of organs, because no center wanted to be closed. Whereas the number of transplants increased every year before the COP, they stabilized for several years and there was no increase [62]. Similarly, publication of the COP for OPOs led to a stabilization of the number of deceased donors recovered, although it too had been increasing in previous years. Like transplant centers, no OPO wanted to be closed if it could not meet the performance measures. By limiting the number of transplants, acceptance of organs, and limiting the number of deceased donors they recovered, both transplant centers and OPOs sought to continue their existence. Thus, regulation may have an adverse effect on some aspects of transplantation. Another issue for regulators is to design performance measures that do not stifle innovation and experimentation. How can transplant centers that want to try new, unproven techniques or therapies not be punished if these innovations turn out to be unsuccessful and lead to worse outcomes? Because of the negative feedback about COPs from transplant centers and OPOs and the adverse effect, the COPs may be having on outcomes and the number of transplants performed; CMS is currently engaging transplant centers and OPOs to revise performance metrics. Hopefully, these revised metrics will remove disincentives and will result in more patients being transplanted and better outcomes.

3. Conclusion

The Ciba Foundation held the first international, interdisciplinary conference on ethical and legal issues in transplantation in March 1966. Some of the issues discussed included the following: definition of death, removal of kidneys from moribund but not yet dead patients, use of living kidney donors, ensuring consent for kidney removal in living donors is voluntary, organ markets, and economic barriers to transplantation [63]. After 50 years, many of these ethical issues remain. How best to organize organ transplantation, to increase the number of organ donors, allocate organs, and regulate transplantation and organ donation are constantly works in progress. Changes try to improve patient access, improve transplantation outcomes, and increase the number of transplants. As long as the shortage of organs continues, as it will for the foreseeable future, there will be ethical challenges to confront. Any changes may solve some ethical problems but are likely to introduce new ones.

Author details

Richard J. Howard* and Danielle L. Cornell

*Address all correspondence to: howarr@lifequest.ufl.edu

LifeQuest Organ Recovery Services, Gainesville, FL, USA

References

[1] Howard RJ, Cornell DL, Cochran L. History of deceased organ donation, transplantation, and organ procurement organizations. Progress in Transplantation 2012; 22: 1.

[2] Howard RJ, Cornell DL, Schold JL. CMS oversight, OPOs and transplant centers and the law of unintended consequences. Clinical Transplantation 2009; 23: 778–783.

[3] Howard RJ. We have an obligation to provide organs for transplantation after we die. American Journal of Transplantation 2006; 6: 1786–1789.

[4] Wijdicks EF, Varelas PN, Gronseth GS, Greer DM. Evidence-based guideline update: Determining brain death in adults. Report of the Quality Standards Subcommittee of the American Academy of Neurology 2010; 74: 1911–1918.

[5] Rady MY, Verheijde JL, McGregor JL. Scientific, legal, and ethical challenges of end-of-life organ procurement in emergency medicine. Resuscitation 2010; 81: 1069–1078.

[6] Bos MA. Ethical and legal issues in non-heart-beating organ donation. Transplantation Proceedings 2005; 37: 574–576.

[7] Veatch RM. Killing by organ procurement: Brain-based death and legal fictions. Journal of Medicine and Philosophy 2010; 35: 154–179.

[8] Miller FG, Truog RD, Brock DW. The dead donor rule: Can it withstand critical scrutiny? Journal of Medicine and Philosophy 2010; 35: 299–312.

[9] Truog R. The price of our illusions and myths about the dead donor rule. Journal of Medical Ethics 2016; 42:318–319.

[10] Middleton PF, Duffield M, Lynch SV, et al. Living donor liver transplantation—adult donor outcomes: A systematic review. Liver Transplant. 2006; 12: 24–30.

[11] Segev DL, Muzaale AD, Carro BS, et al. Perioperative mortality and long-term survival following live kidney donation. JAMA 2010; 303: 959–966.

[12] Ibramim HN, Foley R, LiPing T, et al. Long-term consequences of kidney donation. The New England Journal of Medicine 2009; 360: 459–469.

[13] Mjoen G, Hallan S, Hartmann, A, et al. Long-term risks for kidney donors. Kidney International 2014; 86: 162–167.

[14] http://www.srtr.org

[15] Moore FD. New problems for surgery: Drugs that act on the cell nucleus affect the surgeon's work on cancer and on transplantation. Science 1964; 144:388–392.

[16] Delmonico F, Abecassis M, Adams M, et al. Life Organ Donor Consensus Group. Consensus statement on the live organ donor. JAMA 2000; 284: 2919–2926.

[17] Davis CL, Cooper, M. The state of U. S. living kidney donors. Clinical Journal of the American Society of Nephrology 2010; 5: 1873–1880.

[18] Grams ME, Sang Y., Levey AS., et al. Kidney-failure risk projection for the living kidney-donor candidate. The New England Journal of Medicine 2016; 374: 411–421.

[19] Steiner R. Risk appreciation for living kidney donors: Another new subspecialty? American Journal of Transplantation 20034; 4: 694–697.

[20] Testa G. Ethical issues regarding related and nonrelated living organ donors. World Journal of Surgery 2014; 38: 1658–1663.

[21] Dew MA, Jacobs CL, Jowsey SG et al. Guidelines for the psychosocial evaluation of living unrelated kidney donors in the United States. American Journal of Transplantation 2007; 7:1047–1054

[22] Al Sebayel M, Alenazi AM, Sabbagh R, Al Enazi T, Al Bahili H, Elsiesy H. Donor organ shortage crisis: A case study review of a financial incentive-based system. Transplantation Proceedings 2014; 46: 2030–2035.

[23] Peters TG, Fisher MD, Gish RG. Howard RJ. Views of US voters on compensating living kidney donors. JAMA Surgery. 2016; Mar 23. doi: 10.1001/jamasurg.2016.0065. (Epub ahead of print). p. 1–7.

[24] Harmon W, Delmonico F. Payment for kidneys: A government-regulated system is not ethically achievable. Clinical Journal of the American Society of Nephrology. 2016; 1: 1146–1147.

[25] Matas AJ, Hays RE. A regulated system of incentives for living kidney donation: It is time for opposing groups to have a meaningful dialogue! American Journal of Transplantation 2014; 14: 1944–1945.

[26] Matas AJ, Schnitzler M. Payment for living donor (vendor) kidneys: A cost-effectiveness analysis. American Journal of Transplantation 2003; 3: 216–221.

[27] Ghods AJ, Sava S. Iranian model of paid and regulated living-unrelated kidney donation. Clinical Journal of the American Society of Nephrology 2006; 1: 1136–1145.

[28] Jahromi AH, Fry-Revere S, Bastani B. A revised Iranian model of organ donation as an answer to the current organ shortage crisis. Iranian Journal of Kidney Diseases 2015; 9: 354–360.

[29] Salomon DR, Langnas AN, Reed AI, et al. AST/ASTS workshop on increasing organ donation in the United States: Creating an "arc of change" from removing disincentives to testing incentives. American Journal of Transplantation 2015; 15; 1173–1179.

[30] Arnold R, Bartlett S, Bernat J, et al. Financial incentives for cadaver organ donation: An ethical reappraisal. Transplantation 2002; 73: 1361–1367.

[31] Rothman SM, Rothman DJ. The hidden cost of organ sale. American Journal of Transplantation 2006; 2: 1524–1528.

[32] Monaco AP. Rewards for organ donation: The time has come. Kidney International 2006; 69: 955–957.

[33] Veatch RM. Why liberals should accept financial incentives for organ procurement. Kennedy Institute of Ethics Journal 2003; 13: 19–36.

[34] Gill MB, Sade RM. Paying for kidneys: The case against prohibition. Kennedy Institute of Ethics Journal 2002; 12: 17–45.

[35] Caplan A, Coelho DH. The Ethics of Organ Transplants: The Current Debate. 1998. Prometheus Books, Amherst, NY.

[36] http://www.ama-assn.org/amednews/2008/07/07/prsm0707.htm

[37] Delmonico FL, Arnold R, Scheper-Hughes N, et al. Ethical incentives—not payment—for organ donation. NEJM 2002; 346: 2002–2005.

[38] https://optn.transplant.hrsa.gov/data/view-data-reports/national-data/

[39] Klassen DK, Edwards LB, Stewart DE, et al. The OPTN deceased donor potential study: implications for policy and practice. Am J Transplant 2016; 16: 1707–1714.

[40] Howard RJ. We have an obligation to provide organs for transplantation after we die. American Journal of Transplantation 2006; 6: 1786–1789.

[41] Merle J-M. A Kantian argument for a duty to donate one's own organs. A reply to Nicole Gerrand. Journal of Applied Philosophy 2000; 17: 93–101.

[42] Etzioni A. Organ donation. A communitarian approach. Kennedy Institute of Ethics Journal 2003; 13: 1–18.

[43] Menzel PT. The moral duty to contribute and its implications for organ procurement policy. Transplantation Proceedings 1992; 24: 2175–2178.

[44] Howard RJ. How can we increase the number of organ and tissue donors? Journal of the American College of Surgeons 1999; 188:317–327.

[45] Goldberg DS, Trotter JD. The gift that keeps on giving: Increasing donation rates by offering incentives. Am J Transplant 2016; May 25 doi: 10.1111/ajt.13850. (Epub ahead of print), p. 1–2.

[46] Veatch RM, Pitt JB. The myth of presumed consent: Ethical problems in new organ procurement strategies. In. Caplan AL, Coelho DH, eds. The Ethics of Organ Transplants. The Current Debate. 1998. Prometheus Books, Amherst, NY, p. 173.

[47] Miranda B, Fernandez Lucas M, de Felipe C, Naya M, Gonzalez-Posada JM, Matesanz. R. Organ donation in Spain. Nephrology Dialysis Transplantation 1999; 14(Suppl. 3): 15–21.

[48] http://www.ama-assn.org/ama/pub/physician-resources/medical-ethics/code-medical-ethics/opinion2155

[49] Ambagtsheer F, Weimar W. Organ trade: Knowledge, awareness, and nonlegislative responses. Transplantation 2016; 100: 5–6.

[50] Participants in the International Summit on Transplant Tourism and Organ Trafficking Convened by The Transplantation Society and International Society of Nephrology in Istanbul, Turkey, April 30 through May 2, 2008. The Declaration of Istanbul on organ trafficking and transplant tourism. Clin J Am Soc Nephrol. 2008; 3: 1227–1231.

[51] Rothman DJ, Rose E, Awaya T. The Bellagio Task Force report on transplantation, bodily integrity, and the international traffic in organs. Transplant Proceedings 1997; 29: 2739–2745.

[52] Bakdash T, Scheper-Hughes N. Is it ethical for patients with renal disease to purchase kidneys from the world's poor? PLoS Medicine 2006; 3: 1699–1702.

[53] Radcliffe-Richards J, Daar AS, Guttman RD, et al. The case for allowing kidney sales. Lancet 1998; 351: 1950–1952.

[54] https://optn.transplant.hrsa.gov/resources/ethics

[55] Chumfong I, Brown D, Keume J, et al. Distributing a limited resource: Ethical allocation of deceased donor kidneys. Surgery 2014; 156: 198–203.

[56] Batabyal P, Chapman JR, Wong G, Craig JC, Tong A. Clinical practice guidelines on wait-listing for kidney transplantation: Consistent and equitable? Transplantation 2012; 94: 703–713.

[57] Caplan A. Bioethics of organ transplantation. Cold Spring Harbor Perspectives in Medicine 2014; 4: 1–7.

[58] Howard RJ. The challenging triangle: Balancing outcomes, transplant numbers, and costs. American Journal of Transplantation 2007; 7(11): 2443–2445.

[59] Heilman RL, Mathur A, Smith ML, et al. Increasing the utilization of kidneys from unconventional and higher risk deceased donors. Am J Transplant. 2016 June 14. doi: 10.1111/ajt.13867, p. 1–7.

[60] Centers for Medicare & Medicaid Services (CMS), HHS. Medicare and Medicaid Programs; Conditions for coverage for organ procurement organizations (OPOs). Final Rule (42 CFR 413, 441). Fed Registry 2006; 71: 30982.

[61] Centers for Medicare & Medicaid Services (CMS), HHS. Medicare program; hospital conditions of participation: Requirements for approval and re-approval of transplant centers to perform organ transplants. Final Rule (42 CFR 405, 482, 488, 498). Fed Registry 2007; 72: 15198.

[62] Howard RJ, Cornell DL, Schold JL. CMS oversight, OPOs and transplant centers and the law of unintended consequences. Clinical Transplantation 2009; 23: 778.

[63] Ross L, Thistlethwaite JR. The 1966 Ciba Foundation symposium on transplantation ethics: 50 years later. Transplantation 2016; 100: 1191.

Chapter 8

In Whose Best Interests? Critiquing the "Family-as-Unit" Myth in Pediatric Ethics

Joseph A. Raho

Additional information is available at the end of the chapter

http://dx.doi.org/10.5772/66715

Abstract

In pediatrics, parents are the presumed surrogate decision-makers for their children. Parents are generally obligated to make decisions in the child's *best interest*. When assessing what is in the child's best interests, parents should consider the child's experience of illness, potential for suffering (physical or psychological), and ability to understand and tolerate treatment. Yet, parents may consider a variety of factors other than best interest when making treatment decisions for their children. Moreover, parents may equate the child's best interest with their own (or their family's) and make decisions that, in some situations, will place children at significant risk of serious harm. Clinicians may be reluctant to challenge parents due to a perception that their obligations require treating the family "as a unit." After detailing a case from the author's own practice in clinical ethics, this essay will challenge the view that "family-centered" (as opposed to "patient-centered") care is an appropriate ethical model for pediatric decision-making. Specifically, the physician-patient relationship—or, in this context, the *pediatrician-child relationship*—ought not to be reconceptualized into the *pediatrician-parent-child relationship*, since the latter perspective potentially misidentifies who the patient is and may inadvertently suggest there is warrant for "treating" the family's suffering at the expense of the child's welfare.

Keywords: pediatric bioethics, family-centered care, best interest standard, surrogate decision-making, pediatrician-child relationship

1. Introduction

In the absence of clear direction from the patient at a prior time of capacity, and when the patient's treatment preferences cannot be inferred from knowledge of his or her values, surrogate decision-making will be based on the *best interest standard*. The pediatric setting[1] is no different, where parents are the presumed decision-makers for their children.[2] According to this standard, parents are obligated to make decisions based on factors such as the child's apparent experience of illness, potential for suffering (physical or psychological), and ability to understand and tolerate treatment [1]. Yet, in the clinical setting, parents may be influenced by a variety of factors other than best interest when making treatment decisions for their children [2]. From my own experience, parents may equate the child's best interest with their own (or their family's) and make decisions that, in some situations, will place children at significant risk of serious harm. Clinicians, for their part, may be reluctant to challenge parents due to a perception that their medical obligations in this setting require treating the family "as a unit."

After detailing a case from my own practice in clinical ethics, this essay will challenge the view, held by some bioethicists and clinicians, that "family-centered" (as opposed to "patient-centered") care is an appropriate ethical model for pediatric decision-making. I will argue that the moral basis of pediatric ethics is not qualitatively different than the ethics of medicine in the adult setting. Specifically, the physician-patient relationship—or, in this context, the *pediatrician-child relationship*—ought not to be re-conceptualized into the *pediatrician-parent-child relationship*, since the latter perspective potentially misidentifies who the patient is and may inadvertently suggest there is warrant for "treating" the family's suffering at the expense of the child's welfare. This thesis will be sustained from a reflection on the moral responsibility incumbent on surrogate decision-makers. Concluding remarks will offer practical suggestions for improving decision-making at the bedside.

2. Case[3]

Mrs. Rusin was 30 weeks pregnant with her second child. During a routine obstetric visit, she was found to have placental insufficiency, causing an abnormal fetal heart rate; Mrs. Rusin was rushed to the hospital for emergency delivery. The infant was named Konstantin, after Mrs. Rusin's own father who was still living overseas. Immediately after birth, Konstantin was transferred to the hospital's neonatal intensive care unit (NICU) for prematurity, sepsis,[4] and further heart monitoring. Mrs. Rusin was moved from Labor and Delivery to a regular hospital bed for recovery, where she remained for 2 days. During this time, medical decisions for Konstantin were being authorized by his father in the NICU.

---

[1] Pediatrics is the branch of medicine that focuses on infants, children, and adolescents. Neonatology is a subspecialty of pediatrics that focuses on the medical care of newborn infants.

[2] This presumption could be rebutted if, for example, the parents themselves were incapable of medical decision-making or had conflict of interests.

[3] This is a composite case; the names and details contained within have been de-identified.

[4] Neonatal sepsis refers to "invasive infection, usually bacterial, occurring during the neonatal period." Indications of sepsis include "diminished spontaneous activity, less vigorous sucking, apnea, bradycardia, temperature instability, respiratory distress, vomiting, diarrhea, abdominal distension, jitteriness, seizures, and jaundice" [3].

http://dx.doi.org/10.5772/66715

"Everything possible should be done," Mr. Rusin instructed the clinicians. Consequently, when Konstantin's medical condition deteriorated further, cardiopulmonary resuscitation and multiple life-sustaining supports (mechanical ventilation, vasopressors, and artificial nutrition and hydration) were initiated. On day three, Mrs. Rusin was discharged from the hospital; she joined her husband at bedside and agreed that all treatments should be provided. At this early stage in his hospitalization, the cause of Konstantin's ailments and chances for recovery were uncertain; for this reason, the medical team did not think Mr. and Mrs. Rusin's requests were unreasonable.

Several days later, despite maximal ventilatory support, Konstantin did not show any signs of improvement. He could not be weaned from the ventilator and he remained dependent on artificial tube feeds. By day seven, he developed uncontrollable seizures, believed to be due to intracranial hemorrhage. There was concern for significant neurological impairment. Despite efforts to control his seizures, the episodes continued, sometimes multiple times each day.

During one of these acute deteriorations, the bedside nurse could perceive that Konstantin was in significant distress. She related her concern to the parents, but they dismissed it outright: "We only need more time before Konstantin will finally be well enough to come home and meet his brother." The nurse gently responded: "In my experience, infants like Konstantin may be too sick to leave the hospital. He has already suffered some cognitive impairment. If his seizures don't resolve soon, he is at risk for global neurological damage. If that happens, I believe we should not pursue aggressive measures to sustain his life. He will only continue to suffer." Mrs. Rusin and her husband were indignant, indicating that they would never consent to such treatment limitations, especially those that would result in Konstantin's death. In Mrs. Rusin's words, "We are the decision-makers; we have the right to decide what we do and don't want!"

Another week passed with no clinical improvement. The NICU team recommended to the parents to agree to a "do-not-resuscitate" (DNR) order and not escalate treatment further; Mr. and Mrs. Rusin remained unwilling. At this point, the medical team requested an Ethics Consult. A multidisciplinary team and family meeting were held the next day. In attendance were the attending physician, medical resident, bedside nurse, palliative care team, social worker, chaplain, ethicist, and Mr. and Mrs. Rusin. The purpose of the meeting was to discuss Konstantin's poor prognosis, likelihood for recovery, and to better understand the parents' perspectives regarding goals of care. The attending physician related that, despite numerous efforts, the team remained unable to control Konstantin's seizures; in his opinion, Konstantin had suffered and would continue to be at risk for suffering, significant, irreversible neurological damage. His anticipated prognosis was believed to be 1 year, although a more realistic estimate was 6 months. The ethics consultant explained how, in such cases, the parents are required generally to make decisions in Konstantin's best interests. Mr. and Mrs. Rusin took a moment to digest this new information; they then asked whether it would be possible to take Konstantin home, even if it were anticipated that he would eventually die in that setting. The attending related that Konstantin is permanently ventilator dependent and feeding tube dependent and that it is very unlikely that he would be able to go home. "In such unfortunate circumstances, I would recommend that we simply make Konstantin comfortable. We can remove aggressive life supports and allow him to die with optimal palliative care. This is called *comfort care*."

There was a long, drawn-out pause. Mr. Rusin took a deep breath. He then spoke.

"My wife and I, we ... we simply *need* more time. This is all happening too quickly. Konstantin might turn around. How could we live with ourselves if we give up before giving him a chance? Unless you tell us there is no opportunity for improvement, we want to continue therapy for the next couple of weeks." Although they thought it unlikely that Konstantin would improve to be discharged home, the team reluctantly agreed to continue aggressive efforts for a 2-week trial period. The team did manage to persuade Mr. and Mrs. Rusin to agree to a DNR order.

Konstantin did not improve during the interim—his seizures only intensified further. The nursing staff became increasingly distressed. The attending physician confirmed—after receiving a senior second medical opinion and seeking the guidance from expert consultants—that Konstantin's condition was too fragile for him to be discharged home, even on home hospice. With this updated information, Mr. and Mrs. Rusin better appreciated the magnitude of the situation. Since the option of returning home was definitively closed, the parents were again asked if they would consent to transition Konstantin to full comfort care and allow him to die.

Mrs. Rusin sat down and started crying. In between sobs she related: "My father ... he has always been present during my most difficult decisions, and—and—he's not here." There was a brief pause. "Konstantin's life support can't be turned off without Daddy by my side. I ... I wouldn't be able to cope." Mr. Rusin put his arms around his wife and addressed the team: "This has been a whirlwind for us. In a couple of days, our first son celebrates his fourth birthday. As parents, we also need to protect him, too. I can't allow him to be scarred for years because his birthday coincided with his younger brother's death." To this, the attending responded: "I clearly recognize your profound need for support during this difficult time. How about we wait several days until after your son's birthday? That should give enough time for Mrs. Rusin's father to arrive."

What the attending did not realize, however, is that Mrs. Rusin's father lived abroad and did not have a visa for entry to the United States. Obtaining an "emergency" visa could take upwards of 2 weeks. Mrs. Rusin clarified this. She stated, "As soon as he's here with us, I will allow the supports to be turned off. Until then, I cannot do it."

All involved and agreed that the best course of action would be to transition Konstantin to comfort care as soon as possible. However, the Rusin family was also in a time of great emotional need. As one team member put it, "We always strive to provide family-centered care. This means we treat the whole family. Their needs must be considered, too." Although the NICU team was deeply disturbed with the prospect of waiting 2 weeks before transitioning Konstantin to comfort care, the attending physician eventually acceded to the parents' request.

Konstantin continued to seize. Two weeks later, once Mrs. Rusin's father arrived, life supports were discontinued. Konstantin died soon thereafter.

http://dx.doi.org/10.5772/66715

3. Case reflections

Deciding for others in any medical context is complex. The pediatric setting is no different. Below, we discuss a number of complexities that emerged during this case.

Clinicians are entrusted with significant responsibility for their patients' well-being. Yet, clinical judgment is inherently probabilistic and practiced amidst uncertainty. Prognostication, therefore, presents unique challenges to the treating team. Clarifying those treatment modalities that may offer some realistic potential for benefitting Konstantin will be directly related to his anticipated prognosis. However, because his survival and anticipated quality of survival are initially difficult to characterize, ascertaining what would be in his best interests is not immediately clear. Should he remain in the NICU for another 2 weeks, in order to gain more certainty about his clinical trajectory? Or, would it be better to withdraw aggressive treatment, since his condition has not improved and his clinicians do not suppose this is likely to change? When communicating such options to Konstantin's parents—a process which presents its own obstacles—Mr. and Mrs. Rusin may be skeptical: "How can be we sure that he will *not* improve? Is it not the case that physicians are sometimes mistaken?" Assessments of probability and risk are value-laden evaluations; it should not surprise us when parents and clinicians disagree regarding how to weigh each variable. Are parents to blame when, in the absence of clinical certainty, they insist that "everything possible" be done to try to save their child?

Families experience significant stress, fear, and uncertainty during their child's stay in the NICU. It is a foreign setting [4]. There are new sounds—beeping monitors, infants in distress, and the constant (albeit brief) encounters with the large clinical team (attending physician, medical fellow and resident, nurses, consulting specialists, palliative care experts, social workers, chaplains, and ethicist). Parents are often unprepared, overwhelmed, shocked, and may experience hopelessness and despair [5]. In one study, some of the highest reported stressors identified by parents included "feeling helpless about how to help my baby during this time"; "seeing my baby stop breathing"; and noticing that "my baby seems to be in pain" [5]. The subjective experience of parents may be likened to an "emotional roller-coaster" [6]. It is common for parents to find the experience stressful and disruptive [7] to their everyday routine, and the routines of their children at home. This may be the first time that parents are asked to make "life-or-death" decisions under significant pressure. Deciding between a "certain" death of a child—by placing limitations on the aggressiveness of interventions—versus continuing treatment in the hope that it "may eventually work"—is terrifying for families. Strong emotions come into play, including "the belief that the death of a child violates the natural life cycle, parental guilt, crushed familial expectations, and the sudden nature of the illness or injury" [2]. Parents (and their families) will live with the consequences of their decision for months, even years. For example, Jones and colleagues note that "Bereaved parents are at risk for anxiety, depression, suicidal ideation, prolonged grief, decreased quality of life, relationship struggles, and social decline" [8]. In addition to the vulnerability inherent in being a patient, the vulnerability of parents, too, is significant and should be addressed.

Their vulnerability in this case manifested itself on multiple levels. Initially, Mrs. Rusin was a patient, undergoing emergency surgery; she later became separated from her child at a crucial moment.[5] Mr. and Mrs. Rusin also were separated during their first 2 days following the delivery of Konstantin; Mr. Rusin may have felt significant stress by making critical decisions in his wife's absence, and Mrs. Rusin probably felt to some extent abandoned and sequestered from her family. Several days later, Mr. and Mrs. Rusin were told that their child was not improving clinically. Yet, this fact may not have been readily apparent to them. Clinical information between the team and family is asymmetrical. Nonclinicians may see matters differently: Konstantin remains in the NICU, he had been tolerating his tube feeds (regardless of whether or not they are "artificial"), and the ventilator was allowing him to breathe; without these supports, Mr. and Mrs. Rusin's child would certainly die. Is there any alternative in these circumstances? Furthermore, although Konstantin was reported to be seizing, the seizures may have been subclinical and reported only by the bedside nurses. Mr. and Mrs. Rusin may not have registered Konstantin's pain and discomfort and assumed that the doctors were simply "going along" with the nurses. Thus, a disconnect may occur between teams and parents regarding whether or not the infant is in pain [10].

As we have stated, adjusting to the NICU takes time for parents. Yet, during this process, the parents are told by the team about a request for an Ethics Consultation. One of the first questions Mr. and Mrs. Rusin probably had asked themselves is, "Are we (or someone else) acting unethically?" Many clinicians have never heard of "Ethics" as a consulting service; it should not surprise us if families have not either. The family is then requested to participate in yet another interdisciplinary team meeting—only this time, Ethics will be present. "The 'Ethics Police' is coming to set the record straight," I can imagine the parents saying to themselves. Upon arriving at the meeting, a simple fact becomes obvious: Mr. and Mrs. Rusin are outnumbered. Given the multiple members of the primary team (the attending physician, resident, and bedside nurse), consultants (palliative physician), and others (social worker, chaplain, and ethicist), a not entirely unreasonable interpretation would be: "They are all ganging up on us."

During the meeting, Ethics explains that decisions must be made in the infant's "best interest." This would appear to be relatively straightforward, until one wishes to attempt to define what it means to choose in the infant's best interest. Is it better to live with the assistance of life-sustaining treatment or die? This question is challenging enough in the adult setting when the patient in question had expressed some treatment preferences in the past or at least had written them down in an enduring advance directive. Consider, by contrast, the situation in which an adult patient had never expressed his wishes regarding medical treatment, but we know reasonably well what he was like, based on reports from the family regarding his values; in this situation, we can try to "reconstruct" what his wishes are likely to be in the current situation. Of course, there are important epistemological questions that could arise about how we truly know what we think the patient would want now. In contrast to all of this, we have the current situation, in which Mr. and Mrs. Rusin must decide in Konstantin's best interest. And, here, we must acknowledge significant disagreement among members of the team, as well as the family. Let us now move to consider the best interest standard.

[5] For the importance of maternal bonding, see [9].

http://dx.doi.org/10.5772/66715

4. Departing from the best interest standard

The *best interest standard* applies when a patient had never expressed capable preferences with regard to medical treatment and his or her preferences cannot be inferred from the surrogate's knowledge of the patient. For neonates such as Konstantin, this is the default decision-making standard.[6] In assessing what will be in their child's best interests, Mr. and Mrs. Rusin will need to choose in a way that will "promote maximally the good (i.e., well-being)" [11] of Konstantin. More specifically, this standard directs Mr. and Mrs. Rusin "to determine the net benefit for the patient of each option, assigning different weights to the options to reflect the relative importance of the various interests they further or thwart, then subtracting costs of 'disbenefits' from the benefits of each option" [11]. Important considerations when deciding in this context include "the relief of suffering, the preservation or restoration of functioning, and the quality as well as the extent of life sustained" [12].

The best interest standard has both defenders [13, 14] as well as critics [15–18]. The standard has been criticized for being vague, subjective, and incoherent, and some commentators have proposed alternative criteria for decision-making—for example, the "harm principle" [16], the "not unreasonable" standard [18], and the "basic interests" standard [19]. Rather than entering into this specific debate, I wish here to reflect more closely on Konstantin's case, and ask why, although his best interests were initially appealed to, decisions were ultimately made that were not congruent with those interests. Mrs. Rusin's emotional distress and her older son's anticipated distress were considered (either implicitly or explicitly) by the team to be morally relevant factors in the decision. Such "family interests" would appear to have been validated by the team (however reluctantly) to the detriment of Konstantin's own well-being. Konstantin, who continued to seize, had a protracted and intensified dying process. Why was this allowed to occur?

The reasons, I would submit, are subtle. Mr. and Mrs. Rusin were directed to base their decisions on Konstantin's best interest; neither they, nor the team, were in vocal opposition with that standard. Moreover, neither the team nor the family went so far as to argue that the family's interests, *independent* of Konstantin's, were ethically relevant (a view sometimes advanced in the bioethics literature). Instead, the family based their decision on Konstantin's best interest; only later did they request that the clinicians delay the implementation of that plan of care. Thus, once the decision had been made, its practical implementation had to be agreed upon by all parties. And it is here that the team was in a tough position: either accept the parents' request or risk that the parents will demand that full aggressive treatments continue indefinitely. From an ethical perspective, however, the overall plan of care and the therapeutic means to achieve that plan are not the only relevant aspects; the practical implementation of that plan must also pass ethical muster. It is not uncommon for families to ask for brief extensions of life-sustaining treatment so that a geographically distant loved-one may be able to say final good-byes. Rarer still would be those requests by a family to continue aggressive life-sustaining treatments for a week for the same reasons. Why, then, were

[6] This standard will also be the default decision-making standard in emergency settings for all patients.

Konstantin's treatments permitted to continue for 2 weeks? What made the parents' request appear reasonable?

My suspicion is threefold. First, the team wanted to support the parents during this devastating time, while also ensuring that an appropriate plan of care would eventually take place. Decision-making is a process, and these decisions take time for families to feel comfortable in arriving at a mutually agreed upon plan of care. Second, it is quite natural (and human) to wish to avoid conflict. Disagreement, itself, is time-consuming, stressful, and unpleasant. When an alliance between family and team is most needed, putting pressure on the family would create significant tension. Seeking to avoid conflict in the clinical setting is common. One study found that physicians may acquiesce with families "in order to eliminate a disagreement" [20]. Walking the fine line with Mr. and Mrs. Rusin left the team in a compromise position. Third, and most significantly from my perspective, the parents' request went unchallenged due to a prevalent, albeit misguided, interpretation of what it means to provide *family-centered care*. We will focus on this aspect in the remainder of this article. I will argue that an invalid inference was made in assuming that the provision of family-centered care entails giving ethical deference to the family's interests (as somehow independent of the infant's best interests). This does not follow. As I detail below, an *approach* to providing care is not a *standard of decision-making*. This is not to say, however, that teasing these apart in an actual case is never without its challenges.

5. From *family-centered care* to the *family's interest* in care

In the past, hospitalized children were cared for almost exclusively by health care professionals; parental visitation was significantly (if not completely) restricted [21]. The standard of decision-making was clearly paternalistic: doctors decided what was in the child's best interest. This began to change once research began to document the serious consequences that occur to the child—emotional, psychological, and developmental—if separated from the mother for a significant period of time [21]. The rise of family-centered care, I would suggest, should be understood against this backdrop. It is not insignificant that the early stages in the development of family-centered care generally took place amidst other major rights movements (e.g., civil rights, patients' rights, women's rights)—movements that in many ways challenged the traditional paradigm in which medicine had hitherto been practiced. Instead of a standard that gave deference to the physician's perspective, family-centered care placed primacy on the parents' determination of what is in the child's best interest. Who was in authority to decide in these cases changed, whereas the decision-making standard remained the same. Families, notably parents, became the default surrogate decision-makers for their children.

As simple as this shift was, it would appear to have changed yet again. Whereas family-centered care originally meant that the family's interpretation of what would be in the child's best interests was decisive, advocates of a family-centered approach would appear to be arguing for something different: the family-centered approach means that the family's interests should be factored in treatment decisions. The family's interests, some might hold, necessarily include the child's interests. What does a commitment to family-centered care really mean?

http://dx.doi.org/10.5772/66715

Family-centered care is the dominant approach in pediatrics today.[7] Family-centered care starts from the premise that families have many important roles—for example, safeguarding the well-being of its members—and translates that idea to the medical setting. Partnerships and collaboration among families, patients, and healthcare providers are sought, and various kinds of support (emotional, social, and developmental) are considered "integral components of health care" [24]. Such a partnership, according to the Institute for Patient- and Family-Centered Care, "redefines the relationships in health care" [24]. One might be tempted to speak here of the "patient-parent-pediatrician relationship," to borrow a term from Lantos [25], as though the parents, too, were patients.

Family-centered care, as articulated by a multidisciplinary task force and summarized by Harrison [21], recognizes the following. Whereas medical teams change continually, the child's family remains. Families' diversity (e.g., cultural, ethnic, racial, spiritual, social, economic, and educational) should be recognized and honored. Families have different ways of coping, which means various targeted supports (e.g., developmental, educational, emotional, and financial) should be provided to assist them. And, finally, families and their children "possess a wide range of strengths, concerns, emotions, and aspirations beyond their need for specialized health and developmental services in support" [21].

In my clinical experience, a commitment to family-centered care means in practice that families are given more opportunities to participate actively in their child's care while admitted to hospital. For example, parents may be allowed to sleep in the same room with their child overnight; they participate on daily multidisciplinary rounds; and clinicians seek their input at the bedside. If a child is to be discharged home, teams may teach families how to deliver care in that setting. Families, too, may have medical needs. If a child were to die, bereavement support is available to families.

Family-team partnership makes a lot of sense. Seriously ill children may be admitted frequently to the hospital, and parents and families are the "continuity" caregivers for their child. Young children are literally voiceless; their families must be their advocates, safeguarding the child's interests. Instilling firm partnerships with clinicians makes it more likely that the child will be provided excellent care within and beyond the hospital. As related by the American Academy of Pediatrics, this approach has been shown to "improve the patient's and family's experience, increase patient and family satisfaction, build on child and family strengths, increase professional satisfaction, decrease health care costs, and lead to more effective use of health care resources" [23].

Finally, we must bear in mind that families have a number of important functions, such as preserving the child's life, forging a child's identity, socializing children, nurturing children, and nursing children when they are ill [26]. Furthermore, families also have inherent value; they are "places of love"; "where lives are shared"; where "family members encumber their children with 'thick' conceptions of the good" [26]. In light of the functions and inherent value of families, one might argue that the family's interests (medical or nonmedical) should be taken into account when making treatment decisions for their kin. That is,

[7] It has been endorsed by the American Academy of Pediatrics [22, 23], for example.

one might argue that family-centered care requires that clinicians take into consideration the legitimate interests of families in addition to the medical interests of patients (who happen to be family members). We will consider, and ultimately reject, this suggestion in the next section.

6. The "family-as-unit" myth in pediatrics

In the bioethics literature, many authors—see, for example, [26–29]—contend that the family's interests ought to be given serious consideration when treatment decisions are being made for their sick children. This is for a variety of reasons. First, a family is a "unit." Individual members of that unit will be affected by whatever decisions (medical or otherwise) are made. As Lindemann Nelson has written, "The care of an acutely ill child requires the family to channel many of its resources toward a single member: an arrangement that can usually be sustained for a while but that cannot continue indefinitely while the other members do without. Illness disrupts ordinary familial functions and, if it is serious enough, threatens to break the family altogether" [26]. Second, the lives of family members are intertwined with one another. What happens to one person matters to the others. "There is no way to detach the lives of patients from the lives of those who are close to them," writes Hardwig. "Indeed, the intertwining of lives is part of the very meaning of closeness" [27]. Put differently, family members are "stuck with each other" [30]. Third, families encompass an assortment of interests, both individual and collective. Each family member has unique projects, desires and hopes; families also promote shared aims. At times, the interests of some members will be more relevant to decision-making than others: "the interests of some members sometimes give way to the interests of others, or to the interests of the family as a whole" [28]. Fourth, to be a member of a family means that you are not entirely free to choose as you please [27]; being a member of a family creates responsibilities and obligations. The family may limit the autonomy of its members. Finally, many health care professionals—ethicists included—sometimes act as though the *medical interests* of patients were the only aspect that mattered for the purposes of decision-making. But for families, this is not always true. As Hardwig has argued, "Even life or death is not always the most important consideration. [...] We must beware of the power of the medical context to subordinate all other interests to medical interests. Sometimes non-medical interests of nonpatients morally ought to take precedence over medical interests of patients" [31].

In light of these points, one of two alternative conclusions is plausible: either (a) the family's interests ought to be given consideration when parents make decisions for their child; or, alternatively, (b) parents should choose the option that will promote the family's overall interest, since the child's best interest is intimately connected with the family's interest. I will argue that although consideration should be given to the family's interests, pediatricians are morally obligated to factor only the child's best interests in decision-making. Family interests, I will argue, are generally *secondary* to the medical interests of the patient. I therefore reject (b).

A myth is "a widely held but false belief or idea" (OED). All myths contain elements of truth. Neonates like Konstantin are part of a family; the Rusin family is a group of individuals that

http://dx.doi.org/10.5772/66715

forms a unit; and the life of each of its members matters to the group. Moreover, Mr. and Mrs. Rusin likely have shared communal goals and a vision of the good of family life. One would expect Konstantin and his brother to be inculcated in this ethos. But why should the family's interests matter? What is it about the family's interests that are morally relevant? Is it the fact that long hospitalizations will drain the family's emotional and financial reserves, such that Mr. and Mrs. Rusin would not be able to provide adequately for Konstantin's brother and for each other? These concerns are not insignificant, and clinicians should explore with the family these sources of stress. Nevertheless, what is it about the family's interests that are morally relevant for the pediatrician? Perhaps the family's interests are important when considering general ethics; how might these interests apply to medical ethics? We must remember, although family-centered care is the dominant approach in pediatrics, family-centered care is not a standard of decision-making. It is a myth to think otherwise.

Although being a part of a family generally creates responsibility, the extent of this responsibility will depend on one's *role* within the family. Mr. and Mrs. Rusin have primary responsibilities toward each other, and to their children; very young children do not (at least not initially). Children gradually will acquire such obligations commensurate with their level of maturity. Being a family member at times will require sacrifices of one's interests to the good of the family. Yet, whereas Mr. and Mrs. Rusin may decide to sacrifice their individual interests to the greater good of their family, this expectation cannot be made of Konstantin and his brother (although, presumably, it could in the future). And since Konstantin is currently incompetent, Mr. and Mrs. Rusin may not decide to sacrifice his interests for the greater good of the family: parents may not use their progeny to further their own (or their family's) ends.

Family members have responsibilities to one another. Yet even this is a matter of degree. For example, before making a major career change or relocating his family across country, Mr. Rusin should consult with his family—primarily, his wife.[8] He would be a bad parent as well as a bad husband if he did otherwise. After speaking with his wife, Mr. Rusin may decide that the additional salary he could make at a new job across country would not be the right decision for his family at this time, taking everyone's interest into account. In deciding not to uproot and relocate his family, Mr. Rusin would be sacrificing some of his personal interests—namely, his career-furthering interests—to the greater family good. We expect families to do this since they generally aim to further the collective interests of their members; the life they create is a life shared together. But this does not apply to Konstantin and his brother; they do not have duties analogous to their parents at this time—although they will one day. And yet, whereas very young children cannot be made martyrs for their parents' causes, adult patients may choose to sacrifice themselves in this way for their families as a final gesture of love.[9]

[8] This example is modified from Hardwig [27].

[9] A not infrequent example concerns the adult patient who retains decision-making capacity. He may experience significant symptom burden and yet refuse standard of care medical management—for example, pain medications—in order to remain awake and lucid during family visits. Such an arrangement would be ethically acceptable because the patient capably authorized it. By contrast, it would be wrong to force similarly situated incapable patients to suffer, simply because the family requests it.

Although family interests should be taken seriously by clinicians, such interests are generally *secondary* to the medical interests of the patient, which ought to remain primary. This is for several reasons. First, everyone has interests—families, patients, and clinicians—but not all interests are ethically relevant. Patients have an interest in returning to a state of health, and clinicians have an ethical obligation to focus on the health-related interests of the patient. Thus, in the medical setting, it is the medical interests that matter. This is not to deny or diminish the relevant interests of families. However, the primary obligation of the physician is to this patient, in these circumstances, at this moment in time. It does not really matter that the patient happens to be a child. A patient is still a patient. Physicians have primary obligations to their patients based on their role as healers, and only secondary obligations to patients' families.[10] Second, although the lives of family members are interconnected, we must bear in mind that this state of affairs in the Rusin family is unidirectional—from parent to child, not the other way around. It is not a *relationship* until the child can be an active participant in family life. Until that time, Konstantin remains wholly dependent on his family. The parents are in relation to the child, but the child is not actually in relation to his parents in any meaningful sense. Finally, children may not be sacrificed for their family's causes. We routinely override parents when their idiosyncratic beliefs place the child at significant risk of imminent, irreversible harm. Because very young children are incompetent, they cannot consent to sacrifice themselves. This could shift, of course, as the child matures. A 10-year-old child might be able to make some small sacrifices for his family's overall good. But the physician and family should ask for his permission. We cannot expect this of neonates like Konstantin or small children like his brother.

The obvious objection is that, in at least some cases, families will be greatly affected by the treatment decisions they are required to make according to the best interest of their children. The family's communal interests will be at stake. To this, my response is that, in clinical ethics, there may be exceptions to the general rule. All relevant stakeholders will have to consider the exact point of conflict. Imagine that there were two available treatments for Konstantin's seizures. Option A is invasive, causes discomfort, but otherwise is safe, effective, and covered by his parents' insurance. Option B, by contrast, is not invasive, causes no discomfort, and is otherwise safe and effective, but *not* covered by his parents' insurance. All things being equal, Konstantin's best interest is better served by Option B; however, it is not covered by his parents' insurance. If the cost is high, the treatment may not be in his family's overall interest, relative to other ends it could achieve (e.g., education for Konstantin's brother) with the same money. But notice that this is only a nominal difference between Option A and Option B. Both are safe and effective, so it is really a question of how much discomfort the family would be willing for Konstantin to endure and how much money the treatment will take away from the other valid interests of the Rusin family. By contrast, if Option A were known to be ineffective and Option B effective, the Rusin family might need to make further sacrifices. And if Option B were almost prohibitively expensive, but the only treatment known to reverse Konstantin's seizures and prevent significant neurological

[10] However, compare with Jones et al. [8]: "because the family is the greatest influence in the life of the child, and the well-being of the child and family are intrinsically linked, care for the child must include care for the family. Physicians are ethically bound to care for the family along with the child." The basis for this assertion is unclear, as well as its limits.

http://dx.doi.org/10.5772/66715

devastation, the Rusins might have to sacrifice their collective interests in order to obtain it. Thus, families should have some freedom to balance and weigh the competing interests of its family members in coming to a decision. This does not mean, however, that families should be allowed to make decisions that maximize their interests to the detriment of the medical interests of the child. Families should be overruled in such cases.

7. Concluding thoughts and practical considerations

Although often claimed to be a "right" or "entitlement," decision-making for others is primarily a responsibility to be discharged according to widely recognized legal and ethical standards that have been elaborated over the past several decades. When surrogates do not have clear direction from the patient at a prior moment of capacity, and when the patient's treatment preferences cannot be inferred from knowledge of his or her values, surrogate decision-making will be based on the best interest standard. For adults who have been permanently incompetent, this will be the default standard. For very young children who are currently incompetent, this will be the default standard. It is not clear why that standard should change in pediatrics. "Mini adults" are still human beings; patients with different disease types are still patients; and medicine for children and medicine for elderly adults remain, in essence, the same. The moral basis of pediatric medicine is not qualitatively different than the ethics of medicine in the adult setting. For this reason, I would suggest that the *physician-child relationship* in pediatrics should not be re-conceptualized into the *physician-parent-child relationship*. The latter perspective misidentifies who the patient is and may inadvertently suggest there is warrant for treating the family's suffering at the expense of the child's welfare, as happened in the case of Konstantin.

In light of this analysis, I offer the following three general rules for helping clinical teams to decide how to approach the question of family-centered care in medicine in general.

1. If the patient retains capacity, when making decisions she should be encouraged to take into consideration how this treatment will affect her family. Responsibilities are inherent in family life. She may be convinced that the less-than-optimal treatment should be pursued, so as to maximize the family's interests. Indeed, the capable patient may decide to sacrifice herself to the collective good of the family. This is her right to decide.

2. If the patient is currently incapable of medical decision-making, but there was strong evidence that she would want her family's interests to be primary and her own secondary, clinical teams would have an obligation to follow that previously expressed capable wish. I am thinking here of the patient who might write into an advance directive: "If my continued treatment is so financially expensive, emotionally burdensome, or would require such sacrifices on the part of my family so as to eclipse any real benefit to them or to me, I direct my physicians to refrain from starting life-sustaining treatments and, if such treatment has already been initiated, I direct that it be promptly withdrawn."

3. If a patient had never been capable, one may not infer that she would consent to being sacrificed to the collective good of the family. Incapable patients are not responsible to

their families in the same way that families are to them. Physicians should remind surrogates that decisions in this third category require determining what are in the current best interests of the patient. Surrogates who demand or refuse treatments based on the overall interests of the family (not the patient's best interest) should be overridden.

Based on these general rules, the more compassionate solution in cases such as Konstantin's would be to retain a family-centered paradigm with a patient-focused standard of decision-making.

Acknowledgements

Several of the ideas explored in this paper were initially discussed at length with James Hynds during my clinical ethics fellowship. I remain especially grateful to him. I would like to thank Dan Davis for thoughtful comments on an earlier draft, as well as James Lin, Peter Clark, and Marie-Josée Potvin for helpful suggestions. As always, the responsibility for the content remains with the author.

Author details

Joseph A. Raho

Address all correspondence to: JRaho@mednet.ucla.edu

Clinical Ethicist, Ethics Center, UCLA Health, Los Angeles, CA, USA

References

[1] Nancy Berlinger, Bruce Jennings, and Susan M. Wolf, *Hastings Center Guidelines for Decisions on Life-Sustaining Treatment and Care Near the End of Life* (Oxford, New York: 2013).

[2] David N. Cornfield and Jeffrey P. Kahn, "Decisions about life-sustaining measures in children: in whose best interests? *Acta Paediatrica* (2012) 101(4): 333–336.

[3] Mary T. Caserta, "Neonatal Sepsis," in Merck Manual, available: http://www.merckmanuals.com/professional/pediatrics/infections-in-neonates/neonatal-sepsis (accessed 12/05/16).

[4] Hala M. Obeidat, Elain A. Bond, and Lynn Clark Callister, "The parental experience of having an infant in the newborn intensive care unit," *Journal of Perinatal Education* (2009) 18(3): 23–29.

[5] Cindy Grosik, Denise Snyder, Gerard M. Cleary, Diane M. Breckenridge, and Barbara Tidwell, "Identification of internal and external stressors in parents of newborns in intensive care," *Permanente Journal* (2013) 17(3): 36–41.

http://dx.doi.org/10.5772/66715

[6] Diane Holditch-Davis and Margaret Shandor Miles, "Mothers' stories about their experiences in the neonatal intensive care unit," *Neonatal Network* (2000) 19(3): 13–21.

[7] Margarita Forcada-Guex, Ayala Borghini, Blaise Pierrehumbert, François Ansermet, and Carole Muller-Nix, "Prematurity, maternal posttraumatic stress and consequences on the mother-infant relationship," *Early Human Development* (2011) 87(1): 21–26.

[8] Barbara L. Jones, Nancy Contro, and Kendra D. Koch, "The duty of the physician to care for the family in pediatric palliative care: context, communication, and caring," *Pediatrics* (2014) 133(1): S8–S15.

[9] Ruth Feldman, Aron Weller, James F. Leckman, Jacob Kuint, and Arthur I. Eidelman, "The nature of the mother's tie to her infant: maternal bonding under conditions of proximity, separation, and potential loss," *Journal of Child Psychological Psychiatry* (1999) 40(6): 929–939.

[10] Gay Gale, Linda S. Franck, Susan Kools, and Mary Lynch, "Parents' perceptions of their infant's pain experience in the NICU," *International Journal of Nursing Studies* (2004) 41(1): 51–58.

[11] Allen E. Buchanan and Dan W. Brock, *Deciding for Others: The Ethics of Surrogate Decision Making* (Cambridge, Cambridge University Press: 1990).

[12] President's Commission, *Deciding to Forego Life-Sustaining Treatment: Ethical, Medical, and Legal Issues in Treatment Decisions*, President's Commission for the Study of Ethical Problems in Medicine and Biomedical and Behavioral Research (U.S. Government Printing Office, Washington, D.C.: 1983).

[13] Loretta M. Kopelman, "The best-interests standard as threshold, ideal, and standard of reasonableness," *Journal of Medicine & Philosophy* (1997) 22(3): 271–289.

[14] Loretta M. Kopelman, "The best interests standard for incompetent or incapacitated persons of all ages," *Journal of Law, Medicine, & Ethics* (2007) 35(1): 187–196.

[15] Howard Brody, "In the best interests of…," *Hastings Center Report* (1988) 18(6): 37–40.

[16] Douglas S. Diekema, "Parental refusals of medical treatment: the harm principle as threshold for state intervention," *Theoretical Medicine* (2004) 25(4): 243–264.

[17] Erica K. Salter, "Deciding for a child: a comprehensive analysis of the best interest standard," *Theoretical Medicine and Bioethics* (2012) 33(3): 179–198.

[18] Rosamond Rhodes and Ian R. Holzman, "Is the best interest standard good for pediatrics?" *Pediatrics* (2014) 134(Suppl. 2): S121–S129.

[19] Erica K. Salter, "Resisting the siren call of individualism in pediatric decision-making and the role of relational interests," *Journal of Medicine & Philosophy* (2014) 39(1): 26–40.

[20] Samia A. Hurst, S.C. Hull, G. DuVal, and M. Danis , "How physicians face ethical difficulties: a qualitative analysis," *Journal of Medical Ethics* (2005) 31: 7–14.

[21] Tondi M. Harrison, "Family centered pediatric nursing care: state of the science," *Journal of Pediatric Nursing* (2010) 25(5): 335–345.

[22] American Academy of Pediatrics, Committee on Hospital Care, Institute for Family-Centered Care, Policy Statement, "Family-centered care and the pediatrician's role," *Pediatrics* (2003) 112(3): 691–696.

[23] American Academy of Pediatrics, Committee on Hospital Care and Institute for Patient- and Family-Centered Care, "Patient- and family-centered care and the pediatrician's role," *Pediatrics* (2012) 129(2): 394–404.

[24] http://www.ipfcc.org/faq.html (accessed 12/05/16).

[25] John Lantos, "The patient-parent-pediatrician relationship: everyday ethics in the office," *Pediatrics in Review* (2015) 36(1): 22–30.

[26] Hilde Lindemann Nelson, "Why families matter," *Pediatrics* (2014) 134(Suppl. 2): S97–S103.

[27] John Hardwig, "What about the family?" *Hastings Center Report* (1990) 20(2): 5–10.

[28] James Lindemann Nelson, "Taking families seriously," *Hastings Center Report* (1992) 22(4): 6–12.

[29] Daniel Groll, "Four models of family interests," *Pediatrics* (2014) 134(Suppl. 2): S81–S86.

[30] Lindemann and Lindemann, *The Patient in the Family: An Ethics of Medicine and Families* (Routledge, New York: 1996).

[31] John Hardwig, "The problem of proxies with interests of their own: toward a better theory of proxy decisions," *Journal of Clinical Ethics* (1993) 4(1): 20–27.

Chapter 9

Medical Involvement in Acts of Torture or Degrading Treatment of Human Beings: Forensic and Medical Reflections

Mario Picozzi, Federico Nicoli and Omar Ferrario

Additional information is available at the end of the chapter

http://dx.doi.org/10.5772/65100

Abstract

The following chapter condenses the reflections about the legitimacy of torture, a theme that the authors hope to contribute to in the opening of a debate on this important issue for the future of medicine and for the goals that medicine sets for itself in our time. The topic of this article is relevant to the debate exacerbated by the tragic events of 2001. In the case of capture of terrorists in possession of information regarding imminent attacks, is it permissible to subject them to torture? In what situations and under what conditions is it possible? We will report on the requirements of the critics of the international ban and the justifications for their arguments. We will present the criticisms of those who defend the maintenance of the prohibition of torture. Similarly we will discuss the positions of doctors who are favorable and adverse to participation in procedures of torture.

Keywords: Torture, Medicine, Terrorism, Utilitarianism, Doctors, Attack, Slippery slope principle, Human rights

1. Introduction

The recent controversies regarding the Regeni case (an Italian student who died after being cruelly tortured in Egypt [1]) and the pro-torture statements by a Republican candidate [2] participating in the American presidential primaries show how the issue is more than ever present and relevant. After the September 11 attacks in New York by Osama Bin Laden, as well as those committed by ISIS in London and in Madrid, or even in union with ISIS in Paris, many of our habits as citizens of the Western world in the third millennium have changed. Due to these terrible terrorist attacks, it seems to be licit to ask whether it is permissible to submit captured terrorists who are most certainly in possession of information on future attacks to torture. In addition to reporting the general positions in favor of and against this general situation [3], we are also going to examine the pros and cons of the participation of physicians in these acts, because this implies a conflict between two needs: firstly, that of safeguarding the constitutionally protected value of health and well-being and secondly that of protecting the safety of citizens imposed upon by the war against terrorism. The physicians are not exempt from conflicting theories regarding his/her role as a therapist and his/her position in the state in organizations such as the army; we must also not forget the delicate correlation between medicine and society. We consider it essential to quote a part of the international legislation retarding the definition of torture before examining individual positions about medical participation in torture or inhuman, cruel, and degrading acts.

Faced with this situation, in the fight against so-called urban jihadism represented by "lone wolves" and the "sleeper cells" who are ready to go into action, supervising airports or closing national borders and even dramatically enhancing the supervision of several thousands of sensitive targets may no longer be enough. No place is safe anymore. Therefore the need to fight this new enemy with different weapons could be recognized. Technological high-grade espionage, killing hostile enemies with highly specialized military units or with the use of remote-controlled missiles or drones, continuous controls at borders, and protection of scientific research may also have applications related to national security. The search for possible internal enemies (sleeper cells) with measures that are usually used in the field of counterintelligence, the establishment of detention camps for suspected terrorists, and resorting to special judiciary procedures are all tools that are normally used by the executive international power against an external military threat and that now must be used against an alleged internal enemy.

2. International and supranational legislation

The General Assembly of the United Nations approved the Universal Declaration of Human Rights in 1948. In Article 5 it is stated that "No one shall be subjected to torture or to cruel, inhuman or degrading treatment or punishment" [4].

In this article a specific definition about what could be identified as *inhuman and degrading treatments* and which types of obligations the individual members of the UN have is missing.

An initial definition of torture was provided by the General Assembly's *Declaration on the Protection of All Persons from Being Subjected to Torture and Other Cruel, Inhuman or Degrading Treatment or Punishment* in 1975. Article no 1 defines the torture as "any act by which severe pain or suffering, whether physical or mental, is intentionally inflicted by or at the instigation of a public official on a person for such purposes as obtaining from him or a third person information or confession, punishing him for an act he has committed or is suspected of having committed, or intimidating him or other persons" [5].

Another binding document for the signatory states is the *Convention against Torture and Other Cruel, Inhuman or Degrading Treatment or Punishment* which was put in place on 26 June 1987 (in the international legal language, a convention is binding on the signatory states; this is not true for a declaration). As regards the definition of torture, it replicates the previous declaration of 1975 [6]. In addition, also the *Rome Statute of International Criminal Court* (1998) defines torture replicating that in declaration of 1975 [7].

International legislation has also taken on the issue of the binding nature of the ban; therefore, neither exceptions nor extenuating circumstances subsist. This has been confirmed by many international documents approved by the United Nations General Assembly: article 3 of the *Declaration on the Protection of All Persons from Being Subjected to Torture and Other Cruel, Inhuman or Degrading Treatment or Punishment* of 1975; article 7 of the *International Covenant on Civil and Political Rights* of 1976, in the *Standard Minimum Rules for the Treatment of Prisoners* of 1977, in the *Code of Conduct for Law Enforcement Officials* of 1979; article 6 of the *Body of Principles for the Protection of All Persons under Any Form of Detention or Imprisonment* of 1988; and finally article 2 of the *Convention against Torture and Other Cruel, Inhuman or Degrading Treatment or Punishment*.

On a European level the *Convention for the Protection of Human Rights and Fundamental Freedoms* (*ECHR* [8]) of 1950 not only confirmed the prohibition of torture and humiliating or degrading treatments, but also it established its absoluteness, which cannot be undermined, neither by the law nor by a state of emergency.

In resolution no 690 of the *Declaration on the Police*, the Parliamentary Assembly of the Council of Europe in 1979, in article 3—Appendix 1, Ethics—cites that "summary executions, torture and other forms of inhuman or degrading treatment or punishment remain prohibited in all circumstances. A police officer is under an obligation to disobey or disregard any order or instruction involving such measures" [9].

The binding nature of the prohibition of torture was later confirmed in article 16.3 of the *Copenhagen Conference* by the member states in 1990. Regarding the American continent, both the *American Declaration on the Rights and Duties of Man* of 1948 and the *American Convention on Human Rights* of the 1978 reaffirm the prohibition. The *Inter-American Convention to prevent and punish torture* in article 4 affirmed: "The fact of having acted under orders of a superior shall not provide exemption from the corresponding criminal liability" [10]. From the same convention, article 5 reiterates that "The existence of circumstances such as a state of war, threat of war, state of siege or of emergency, domestic disturbance or strife, suspension of constitutional guarantees, domestic political instability, or other public emergencies or disasters shall not be invoked or admitted as justification for the crime of torture" [10].

3. The definition of torture: some reflections

After examining torture and degrading treatments in terms of international law, we should ascertain their characteristics: the first distinctive feature of torture concerns the active nature of the process or the omission of offering relief from that which causes severe distress or pain, including both the physical and mental consequences of violent actions; the intention of the act, that is, the will of the perpetrator of the crime to intentionally inflict pain and suffering on his/her victims; and the pursuit of a specific purpose: i.e., to obtain information, punish, intimidate, and discriminate. There is also a purpose of torture, which is not explicitly mentioned in the legislation, i.e., to prevent conduct that is contrary to the current political powers that be. In dictatorial regimes, in fact, political opponents are tortured to death in order to send a strong message of deterrence; the role of the public authorities is to represent the perpetrator, i.e., a person with powers authorized by the state: "Enemies are tortured mercilessly for information and then killed. Political opponents are tortured to death to send a powerful deterrent message" [11]. The distinction between torture and inhumane, cruel, humiliating, and degrading treatment concerns primarily the nature of the act, which in turn concerns the intensity of the pain inflicted: "The degree of intensity and the length of such suffering constitute the basic elements of torture, a lot of other relevant factors had to be taken into account. Such as: the nature of ill-treatment inflicted, the means and methods employed, the repetition and duration of such treatment, the age, sex and health condition of the person exposed to it, the likelihood that such treatment might injure the physical, mental and psychological condition of the person exposed and whether the injuries inflicted caused serious consequences for short or long duration are all relevant matters to be considered together and arrive at a conclusion whether torture has been committed" [12]. The analysis of the norms allows for the theory that the definitions of torture and degrading treatment are unique in that certain acts, behaviors, or events may fall within the definition of torture in certain circumstances, or in other situations may not be included, depending on the presence or not of the four elements necessary for the above-presented definition.

Therefore the following statement appears to be relevant: "Torture is not an act in itself, or a specific type of act, but it is the legal qualification of an event or behavior, based on the comprehensive assessment of this event or behavior. […] It is clear that […] the qualification of torture may be easily granted in certain cases. However, in some others, the vulnerability of the victim (age, gender, status, etc), as well as the environment and the cumulative effect of various factors, should be taken into account to determine whether this case amounts to torture or whether it does not reach this ultimate threshold and should be considered as cruel, inhuman or degrading treatment or punishment" [13]. This allows a pragmatic and non-descriptive approach, which at first sight may seem disadvantageous because there is no detailed statement about conduct defined as torture, or degrading or inhuman acts, but which allows including into the definition acts which were not even taken into consideration by the legislature at the time when the norm was written.

By international law, the "mark of Cain" regarding a particular episode of torture would be the utter helplessness of the victim: "Torture, as the most serious violation of the human right to personal integrity and dignity, presupposes a situation where the victim is powerless i.e. is under the total control of another person. […] The decisive criteria for distinguishing torture from cruel, inhuman or degrading treatment may best be understood to be the purpose of the conduct and the powerlessness of the victim, rather than the intensity of the pain or suffering inflicted, as argued by the European Court of Human Rights and many scholars" [14].

4. Favorable positions and the requirements

Those who are in favor of torture are aware of its extreme gravity and limit its use to specific situations, in which a primary objective is at stake—the safety of people. To determine the security of a nation's citizens could be useful recognize the torture, with a previously established legal procedure as "the only way to prevent the bomb from exploding and killing large numbers of civilians" [15]. Thus it becomes ethically legitimate and justified from a legal point of view.

4.1. The interrogation of those who have useful information

Torture must be applied to very specific subjects, who have been ascertained to be in possession of information regarding imminent serious danger. In these cases only the terrorist is to be submitted; few specialists accept the possibility of submitting family members to torture [16]. Most scholars exclude "indirect" measures of this type as they are incompatible with the demands of democracy, which views penal responsibility as strictly personal: "What if it were necessary to torture the suspect's mother or children to get him to divulge the information? What if it took threatening to kill his family, his friends, his entire village? Under a simple-minded quantitative case utilitarianism, anything goes as long as the number of people tortured or killed does not exceed the number that would be saved. This is morality by numbers, unless there are other constraints on what we can properly do. These other constraints can come from rule utilitarianism or other principles of morality, such as the prohibition against deliberately punishing the innocent" [17]. There is a risk to fall down a slippery slope toward the amorality without some limits about the use of torture or other degrading treatments.

In fact, they claim that the application of torture is only against a directly involved subject and against whom they have sufficient evidence that he/she is involved. The subject must be in possession of relevant and necessary information. On this last point, there is a general agreement even though it is unclear how it can be inferred that the person actually possesses that information [16]. This most certainly should be based on the findings from the judicial inquiry [18]. It should be the preponderance of the evidence gathered by investigators which determines the possibility of resorting to torture: "Torture should be permitted where the application of the variables exceeds a threshold level. Once beyond this level, the higher the figure the more severe the forms of torture that are permissible" [19].

4.2. The saving of lives in emergency situations

The saving of lives in emergency situations finds two forms of expression, the most dramatic of which is the ticking bomb scenario (a situation often cited by the defenders of torture). In literature there are variations, but the concept is almost always the fact that a significant number of lives are at stake and thus there is the need to torture the terrorist who knows where the bomb may be: "What do we do? [...] Torturing the terrorist is unconstitutional? Probably. But millions of lives surely outweigh constitutionality. Torture is barbaric?" [20].

This is a theoretical situation in which the existence of a specific society is in danger. In these cases it is not only the loss of human lives the only factor that should be considered but also the great danger that threatens the existence of a State, a community and A culture [21].

The other form of the ticking bomb scenario, which could be called *minor*, involves instead the capture of a terrorist in possession of information regarding imminent attacks, of a nonnuclear nature or with no weapon of mass destruction, in which many people would perish. Unlike before the very existence of a country is not at stake, but only an unspecified considerable loss of life, which is a much less dramatic situation than a nuclear attack.

Instead, in the military setting, the supreme emergency situation is usually defined in relation to the requirements more closely related to possible conflicts also unconventional in progress: "If the unit is a major command, and its being destroyed would cause the war to be lost, this might be considered the supreme emergency addressed earlier" [22].

4.3. Strict rules which are to be observed

The rigid control of the judiciary apparatus regarding the warrant for torture and the public debate should avoid degeneration, giving public legitimacy to something which, if left in secret, could lend itself to serious and unacceptable abuses: "*Off-the-book actions below the radar screen* are antithetical to the theory and practice of democracy" [23].

According to Dershowitz, the use of the presidential mandate should aid in the prevention of the overstepping of the boundaries between torture and abuse, on the basis of the incompatibility of the said abuse with democratic values: "The rights of the suspect would be better protected with a warrant requirement. He would be granted immunity, told that he was now compelled to testify, threatened with imprisonment if he refused to do so, and given the option of providing the requested information. Only if he refused to do what he was legally compelled to do – provide necessary information, which could not incriminate him because of the immunity – would he be threatened with torture" [24].

It should ensure both the judicial protection and the public accountability for the actions of security forces, with the additional protection against abuse, and the *redde rationem* to public opinion, which should be always taken into account by any executive who wants to act in the spirit of the constitution.

4.4. The infliction of intense pain while avoiding permanent damage

Defenders of the use of torture against a terrorist who is uncooperative with the investigation do not go into a lot of detail on what form of pressure is applied. They look at the issue of pain not in legal and ethical terms, but from a pragmatic point of view. Only Dershowitz is quite explicit on the subject. He uses the example of the use of sterile needles under fingernails, based on the maximum sentence principle, i.e., minimum lethality, so as to cause an agonizing pain,

http://dx.doi.org/10.5772/65100

intense and immediate, but without leaving permanent psychological and physical harm: "the use of nonlethal torture [as] a sterilized needle insert under the fingernails to produce unbearable pain without any threat to health or life, or the method used in the film *Marathon Man*, a dental drill through an unanesthetized tooth" [25].

4.5. The justifications

4.5.1. Lesser evil theory

The theories of authors who favor legal and nonlethal torture, such as Dershowitz, are based on currents of thought that could be defined as the "lesser evil": "the question of the lesser evil seems to be the need to choose, as in those situations where the options are, or appear to be, limited and constrained by a great power. The dilemma involves a closed system in which the options available for the choice [...] cannot be questioned" [26]. The first characteristic of the lesser evil theory is that it is a justified exception to the general moral principle that one should never commit evil acts. One must keep in mind one thing, evil is chosen anyway [27]: "Torturing someone to divulge terrorist actions is wrong, no matter what useful information is extracted, and hence no democracy should ever have anything to do with torture. [...] But this style of interrogation [though nonphysical] which would push suspects to the limits of their psychological endurance, would remain a violation of their dignity. It would be a lesser evil than allowing thousands of people to die, but its necessity would not prevent it from remaining wrong" [28]. And we should be aware that someone will have to suffer and that if that person does not pay, it will be someone else. Claiming that it is necessary to restrict the rights of citizens in exchange for security and then placing the responsibility on those whose rights have been restricted are taking a position which is understandable yet hypocritical, as the costs of choosing an evil, albeit minor, should be borne by the person who made that choice, i.e., the responsibility for those choices can only fall on the shoulders of those who have embarked on that specific path.

4.5.2. The rights

The fight against terrorism also poses problems in terms of human and civil rights in emergency situations: "Rights that stand in the way of a regime's survival should be suspended in a time of crisis" [29].

Only in this manner can the government properly manage its monopoly on legitimate violence and overcome the threat to its very existence. Our society is based on the principle of respect for the rights that the individual holds as a person and citizen. Problems arise when the rights of the individuals clash with those of society, and it is necessary to find a balance between them.

A conflict would be between the rights of the few (especially the right to dignity) and the right to life of many people (self-defense). We must never forget the importance of utilizing a proportionality test when it comes to citizens' rights to protection and national security and finding a compromise or a balance of values: "An ethics of balance cannot privilege rights above all, or dignity above all, or public safety above all. [...] They are all important principles – all must be weighed in the balance equally – and nothing trumps" [30].

This argument in favor of legal torture is based on the *extrema ratio* thesis: "In the face of extreme situations, we are quite ready to accept that one should, or even must, sacrifice oneself or *others* for the good of the whole" [31].

A democratic president to whom torture is repugnant is however in the situation of being able to order acts according to a utilitarian calculus where the possible consequences of these actions can be decisive in making us lean toward one behavior rather than another: "If the leader believes he has simply weighed the alternatives and calculated that the consequences of torturing are better than those of not torturing, than he has not committed a crime" [32]. Placed between protecting the rights of an individual and those of society, no utilitarian will find the use of torture to be unacceptable [33].

5. Opposing views

Opposite reasons especially underline how torture is not only ungovernable but also infeasible because it jeopardizes the society in its foundations.

5.1. The requirements

The first objection related to the subject of torture is the terrorist. National and international laws prohibit torturing anyone under any circumstances without distinguishing between different categories of enemies. In this sense, the quibbling about whether the person is a terrorist in possession of information, a fighter but not a belligerent, or an illegal combatant does not help anyone [34].

The law also requires the state and its nationals other obligations under certain dangerous situations, including the protection of the innocent. The term should be understood in the etymological sense, i.e., he who does not injure, not only as a synonym of not guilty. This is one of the thorniest issues of torture [35].

Resorting to the principle of asserting that torturing a captured terrorist will save lives is fraud, and the need for us to force the revelation of a secret that the subject is likely to have does not authorize an act of abuse in such a circumstance. In fact an essential element of self-defense is missing: a direct causal relationship such that the action of defense can probably avoid offense [36].

One of the charges most frequently made against supporters of lifesaver torture in emergency situations regards the argument of the "slippery slope": introducing torture into our system even in exceptional cases would involve a degeneration that eventually would be applied in cases not originally anticipated. It would therefore be impossible to draw a clear line of distinction that would avoid this "slippery slope." Here we merely point out that it represents the most frequent objection to those who support lifesaving torture [37].

http://dx.doi.org/10.5772/65100

Other criticisms concern the inapplicability of the proposed use of torture in emergency situations to everyday life and in particular in "ticking bomb" situations. In such contexts, one wonders how the model proposed by the defenders of lifesaving torture could prove useful. It is true that the model proposed by Dershowitz on paper looks interesting; because of the demand for public legitimacy, one wonders if the reality of the fight against terrorism would prove to be a trump card compared to the current work of law enforcement and judicial authorities, since it would require a capacity of action and reaction that no government agency in the world has [38].

It should be recognized as evil in itself, without the need for laws which prohibit it or not.

5.2. The justifications

The critique that is made against the argument of the *lesser evil* was effectively analyzed by Hanna Arendt: the weakness of the argument has been always obvious—those who choose evil forget too quickly that they are still choosing evil [39].

The argument of the lesser evil has subtle implications. Who presents it does so already taking everything for granted [40]. But the evil aspect is that they make no attempt to put it on a scale where the consequences may be weighed given the presence of increasingly unpredictable factors which in any case escape the possibility of calculation [41]. Those who do defend the thesis of the lesser evil are now enforcing its manipulation of reality; the only possible solution would be to refuse to submit to the imposition of choice.

Torture in exceptional circumstances would not be possible, because this is one of those situations where the exception does not exist. It presents normal men with a dilemma with no alternative: "We can either defend torture by arguing that interrogational torture will provide us with the kind of counterintelligence we need to prevent catastrophic attacks, in which case we need a practice of torture that is effective, or we can defend an absolute ban on torture" [42]. The complaint against what we might call an intermediate approach to the problem is therefore profound: "The moderate position on torture is an abstraction impractical [...] You can call it torture in the dreamland" [42]. The logic of torture in the interrogation process excludes the exception. If, as has been noted by a critic, today's world is made of policies, directives, guidelines, and daily practices and not of emergency measures, there is little to do but admit that the exception invalidates the norm: "we cannot justify even the exceptional case of torture without also justifying a torture culture. If we accept torture we will need torture experts, new instruments of torture, torture research, and a pedagogy of torture" [42].

5.3. The rights

Critics of the anti-torture ban observe international law from the wrong point of view, considering it as a forbidden evil rather than evil itself. The anti-torture laws would be enacted to limit actions that could be permitted in the absence of specific rules. On the contrary, torture should be recognized as evil in itself, which is wrong whether there is an active law which prohibits it or not [43].

In this sense they define those who describe the international anti-torture ban as the expression of a legal archetype: "a rule or positive law that transcends an individual law or statute in that it captures the spirit of an area of law" [44].

Another criticism that opponents of the use of explicit torture put forth is its substantial contradiction with democracy, not to mention the irreparable damage [45] caused to the judicial system. The fundamental incompatibility of torture with constitutional values, according to its opponents, manifests itself in the special violence that is done to a person who in the act of torture is reduced to an object, a thing. In this sense, the comparison made with a rape is perfectly fitting [46].

Those who support the admissibility of the use of torture put forward the argument of the state of exception or *Ausnahmezustand*, a concept which is linked to the doctrine of the famous German jurist Carl Schmitt. In his treatise *Political theology* [47], he defined it as a particular situation in which the executive power, in the face of a threat that endangers its very existence (war between states, civil strife, riots, etc.), assumes total control of all areas of society, suspending rights "*temporarily and in exceptional cases*." This is not the place for an in-depth analysis of the issue. We just keep in mind what has been written on this subject by a philosopher of our day: Giorgio Agamben. In his critique of Carl Schmitt, he emphasizes how the state of exception is not a "pleromatica" state where you have a plus, but rather a "kenomatica" situation that looks like a minus [48].

During an emergency, *which must necessarily be brief*, citizens must take action to repel the threat to their very existence at a time when the law has been suspended, in which what they do has not yet been evaluated from a legal point of view. Despite this, the constitution, civil rights, and possibly also human rights, even if Agamben does not say so explicitly, remain intact. If the president orders an act of torture, the government can and must answer to the law and public opinion. The exception in itself does not justify it. While that is for Dershowitz and his followers, the exception which justifies it in any case, if the required conditions are met, for Agamben, it should be the Court of Justice to determine, regarding the emergency, if it has acted in accordance with the law or not, with possible legal judgments which will differ from case to case.

Not all utilitarians believe torture to be acceptable. Some point out the consequences of the act in its negative terms, primarily the corrosion of the relationship between the government and the people [49].

Utilitarians against torture even quote the slippery slope argument in its two forms—logical and empirical. The first asserts that once certain practices are accepted, from a logical point of view, one is forced to accept others, in that having once embarked on the first step, there are good reasons to stop. So reason would conclude that since the resulting actions are unacceptable, it follows that it is preferable not to take the first step. From a logical point of view, the slope assumes that the acceptance of X contains an implicit justification of Y, even if Y never arises [50]. The empirical wedge shape provides that once certain activities have been accepted, people are implicitly led to accept others, which are much more problematic. The argument refers to what people will do and not to what they have logically committed themselves to

http://dx.doi.org/10.5772/65100

undertake. This involves an empirical prediction of what will happen. It therefore becomes essential to examine context and mindset, because of the psychological influences that they can exert. In logical reasoning we have an orderly progression of similar facts, while in empiricism the pairing of differing facts can also occur. In the case of torture, the empirical version which predicts that if torture was permissible in some circumstances, this will mean an inclination to use it in other situations cannot be justified [51].

6. The doctor and torture

The doctor has a delicate role, and he risks to be forced into a situation of dual loyalty: the society versus the patient and the ethics of war versus the ethics of peace. Is it possible to put together this dual loyalty? Or is it necessary to define priorities to establish tasks and roles in a coherent manner?

6.1. Favorable positions

The justification of the doctor's participation in torture is based on the different roles that the doctor plays according to the contexts in which he/she operates, such as the war against terrorism, where the parameters of medical ethics are not only being differently interpreted but have even been replaced.

When your doctor participates in an interrogation, he is not acting as a doctor, but as a technician. This is the view expressed by the Bush administration and the Pentagon regarding the scandals of Guantanamo and Abu Ghraib [52]. Since he/she is not operating as a health professional, but as a member of an organization like the Army, the doctor-patient relationship does not apply, and medical ethics traditional strongholds such as confidentiality, beneficence, and non-maleficence are not binding. According to this view, their medical degree is a mere certification of technical knowledge and not a sacred vow [53].

Medical ethics in times of war and in times of peace are different. This is the thesis of M. Gross [54]. Here we offer a brief summary for reasons of space. The principles that guide our ability to make decisions in bioethics in ordinary clinical settings are absent when we are at war.

Considering the principle of self-determination, at least for military personnel, it takes a less incisive value than in an ordinary context. Usually we cannot fight compulsory vaccination by law; a soldier who has to leave for a conflict where the use of biological weapons may be expected does not have the same opportunity [55]. Bioethical principles give way to those governing the contemporary *jus in bello*: government motives, proportionality, the doctrine of double effect. On the one hand, the "government motives" require us to act in ways that bring benefits through its conduct to the greater number of its countrymen, considering the collective interest and not the utility to the single individual: "State interest […] transcend aggregate utility to include indivisible collective interest – that is, *super*aggregate utility sometimes defined as *reason of state* or a *way of life*" [56]. On the other hand, military necessity, the fact that "government motives" are a product of times of war, plays a vital role in determining the choices during conflict: "Military necessity functions within particular moral parameters that limit the ends and means of war. Nevertheless, military necessity retains a distinct if not superior status, if only because most individuals gladly weigh the interests of state above their own" [57].

Following Gross' position, the responsibility of the physician to the entire social body is the central issue. The use of torture requires not only the presence of the doctor but also his consent. *He is not just a performer who obeys external orders; the doctor is an active part in decision-making and must defend his position*. In war, according to this interpretation, the traditional principles of ethics change: the principle of beneficence requires the presence of a doctor during the interrogation as a sort of puppet; thus, the suspect being interrogated could be at risk of harm [58]. Regarding the principle of non-maleficence, one author put it this way: "Physicians must not revive interrogates just so that the interrogates can be tortured even more, ultimately being made the worse off for the physician intervention. [...] First, at least in some cases, torture would not be worse than death. Therefore, resuscitating someone merely so she or he can face more torture does not *necessarily* violate the principle of non maleficence *if*, absent resuscitation, she or he would have been even worse off (e.g., dead)" [59].

6.2. Contrary positions

The military doctor is always and above all a doctor, so much so that the international legislation specifies the particular function of belonging to the army in its entirety as an organization of the state but not as a professional corps. According to the latest Geneva Convention, they are to be considered as administrative personnel and combatants and only exceptionally may resort to violence and only to defend themselves or their patients. Assertions reported on the alleged neutrality of the doctor, on his being a mere technician during interrogation, that is, an "interrogator who is medically trained," are categorically refuted by the fact that the presence of health personnel is explicitly required to counter the onset of medical problems: a cardiopulmonary arrest one or laryngeal spasm during a session of water boarding [60].

The defenders of the governmental thesis seem to forget that the doctor, although effectively working for state institutions, must act so as to preserve his/her autonomy and professional independence [61].

Even the thesis of Gross whereby medical ethics in wartime is different from that in time of peace is not without its critics: in fact, according to some authors, neglecting the great importance of human rights and forgetting that beyond the contingent situation medical ethics remain are the same thing.

If respect for human rights is a necessary condition to the decency of a country's political institutions and legal order [62], if torture is a violation of human rights, and if violations of human rights are completely prohibited by international law, then no state or individual can justifiably use or contribute to torture [62].

An approach based on the military doctor as the protector of the patient's compliance constraints and human rights would present a double advantage: not only would he/she be little

http://dx.doi.org/10.5772/65100

influenced by arguments like "necessities of war," but it would provide greater protection against the possibility of abuse [63].

In fact, during court martials a line of defense based on the absolute prohibition of torture would help a military doctor to justify his conduct if he disobeyed an order which involves a violation of international treaties on human rights that his country has signed and that, as a citizen and member of the military, he is bound to observe. Such an order would be unlawful, and his disobedience would be perfectly in line with the provisions of all military justice codes and the living law [64].

Medical participation in torture involves the ordinary breaking of trust that people have for the figure of the doctor and the loss of his integrity as a professional and as a person.

In fact, the silence of the doctors in Guantanamo and Abu Ghraib has had many consequences including the loss of confidence in the prison doctors as defenders of human rights [65]. The participation in medical torture would thus result in the destruction of the professional figure of the doctor and his own personal integrity [66].

What happened in the two US prisons might be the worst demonstration of the empirical form of the slippery slope. Critics emphasize that the physicians' behavior triggered a series of negative events empirically and conferred an aura of legitimacy to abuses, worsening the de facto interrogation situation which the doctor, through his presence, ought to have protected.

Torture therefore shows a tendency to self-expansiveness; that is, it would tend to go easily beyond the limits imposed on it. That fact in itself would make any system built to control it inapplicable. Introducing medical participation in the process would mean setting up ad hoc graduate schools, with instructors, practice tests, and guinea pigs, in order to improve their performance and reduce inefficiency. We live in a world of guidelines and procedures, and in this context, what was established as an exception to the rule would become routine.

7. Conclusion

If torture results in the utter helplessness of the victim, who has lost all capacity of self-determination, the violation of his/her will which often results in a profound splitting of the personality, then the victim becomes an object, a "safe" which is simple to crack.

The mark of Cain of torture, as specified by the UN Special Rapporteur, should be sought in the *utter helplessness of the victim who has lost both his self-determination and dignity* [14]. This definition should be kept in mind in any debate on torture.

If the exceptional nature of the alleged threat "ticking bomb" scenario, apart from the discussions on its alleged unrealistic and excessive nature, entails that evil is still committed, whether human dignity constitutes an absolute principle not comparable with others results in the inability to carry out a balancing of principles though the concept of the lesser gravity of torture with respect to death is absurd, because torture is living and then dying and is worse than death alone [67]: a similar picture appears more in line with those who consider torture evil in itself and always as a legally punishable conduct even if, for a specific and contingent situation, it is carried out in a country with an approved ad hoc law.

One must however consider that the allegations of torture by supporters of saving lives about its usefulness are more similar to anecdotal than anything statistically credible. Of course, this does not mean that there are no cases where torture has been effective, but that it is not possible to use even proper statistical analysis of the matter to support the use of torture under any conditions.

If it is true that torture can neither be restricted nor professionally used [68] and that it corrupts and erodes any professionalism [69] since it constitutes actual acts of abuse, then any attempts to ask supporters of strict regulations of the phenomenon should be considered a chimerical illusion.

The focal point of the discussion lies in the image of state authority that emerges from the pages of the critics of the international ban. If torture is the utter helplessness of the victim, who has lost all ability to self-determination, the rape of his will which may cause irreparable personality disorder, where the citizen becomes an object, a simple safe to crack, then would such a Weltanschauung strengthen or weaken the values of democracy? Should the chief executive respect the constitution, or would it not distort and betray in the name of national security?

In the wake of the ideas of authors like Ignatieff, we believe we should clearly reject torture and admit those forms of coercive pressure, although relatively tough, which in any case interpret the right of the state to use violence as a means to ensure the security of citizens.

Nothing prohibits a government from establishing in peacetime the measures and procedures to be performed in specific exceptional situations of threat, all in compliance with national and international regulations, the prohibition of torture and degrading treatment included.

Only then can the judicial system and the public debate over adherence to established measures with punishment for violations, i.e., pipelines that have exceeded the limits of the terms, decide, once the emergency has passed. In this sense, the exception does not invalidate the general rule because it has not been betrayed, but rather strengthens it, adapting it to a specific context.

Based on what we have said so far, medical participation in torture qualifies not only as a betrayal of the principles of the rule of law and democracy but as a deep "vulnus ethos" of every profession, not only healthcare [70], and thereby provides evidence that the key word with respect to these activities is one and only one: opposition [71]. We believe that the ban should be maintained, and torture and humiliating treatment which are inhumane and degrading should remain prohibited unconditionally, in accordance with most of the literature on the issue.

To strengthen the ban, we align ourselves with the proposal made by some US colleagues for the establishment of an international medical tribunal. This would offset the inherent weaknesses noted in scientific and trade associations and orders, which have proved largely unprepared to deal with the scandal.

http://dx.doi.org/10.5772/65100

An international tribunal would enjoy the support and approval of the UN and would be free from local influences as it would be composed of internationally renowned judges, making it independent of any authorities and political or philosophical thought, and it would have the authority to "investigate allegations of professional misconduct and condemn professionals found guilty of violating norms of professional responsibility" [72]. Albeit with no possibility of imposing criminal penalties, its own power to publicly accuse a culprit of acts of abuse would be a powerful deterrent to grossly unethical conduct through naming and shaming or identify and make the subject of shame.

This court would draw up a code of international conduct and would have the advantage of being able to revoke the professional license of doctors who violate it without those legal restrictions by which national orders can be restricted.

If, for example, a national order cannot sanction the participation of a physician in torture or other acts of abuse because they conform to local law, the court may instead prosecute and convict him and revoke his right to practice that profession whose values have been betrayed. All this without forgetting another important task regarding prevention: the education of future generations of physicians to "recognize torture, treat it, and Report it" [67]. The focus on the education of the younger generation would help reduce the number of doctors who are unanimously considered to be at risk of complicity in torture.

Author details

Mario Picozzi[1], Federico Nicoli[1,2*] and Omar Ferrario[1]

*Address all correspondence to: federico.nicoli82@gmail.com

1 Center for Clinical Ethics, Biotechnology and Life Sciences Department, Insubria University, Varese, Italy

2 Clinical Ethics Service at "Domus Salutis" Clinic, Brescia, Italy

References

[1] Kirchgaessner S. Why was he killed? Brutal death of Italian student in Egypt confounds experts [Internet]. 2016. Available from: https://www.theguardian.com/world/2016/feb/24/why-was-he-killed-brutal-death-of-italian-student-in-egypt-confounds-experts [Accessed: 2016-07-25].

[2] Berenson T. Donald Trump Defends Torture at Republican Debate [Internet]. 2016. Available from: http://time.com/4247397/donald-trump-waterboarding-torture/ [Accessed: 2016-07-25].

[3] Callahan D. editor. The goals of medicine. Setting new priorities. Hastings Center Report, Special supplement, 1996. S14.

[4] United Nations General Assembly. The Universal Declaration of Human Rights [Internet]. 1948. Available from: http://www.un.org/en/universal-declaration-human-rights/ [Accessed: 2016-07-25].

[5] United Nations General Assembly. The universal declaration on the protection of all persons from being subjected to torture and other cruel, inhuman or degrading treatment or punishment [Internet]. 1975. Available from: http://www.ohchr.org/EN/ProfessionalInterest/Pages/DeclarationTorture.aspx [Accessed: 2016-07-25].

[6] United Nations General Assembly. Convention against torture and other cruel, inhuman or degrading treatment or punishment [Internet]. 1984. Available from: http://www.ohchr.org/EN/ProfessionalInterest/Pages/CAT.aspx [Accessed: 2016-07-25].

[7] International Criminal Court. Rome Statute of the International Criminal Court [Internet]. 1998. Available from: https://childrenandarmedconflict.un.org/keydocuments/english/romestatuteofthe7.html [Accessed: 2016-07-25].

[8] Council of Europe. Convention for the Protection of Human Rights and Fundamental Freedoms [Internet]. 1950. Available from: http://hrlibrary.umn.edu/instree/z17euroco.html [Accessed: 2016-07-25].

[9] Parliamentary Assembly of the Council of Europe. Declaration on the Police, resolution n.690 [Internet]. 1979. Available from: http://www.assembly.coe.int/nw/xml/XRef/Xref-XML2HTML-en.asp?fileid=16101&lang=en [Accessed: 2016-07-25].

[10] Organization of American States. Inter-American Convention to Prevent and Punish Torture [Internet]. 1985. Available from: http://www.oas.org/juridico/english/treaties/a-51.html [Accessed: 2016-07-25].

[11] Dershowitz A.M. Why terrorism works. Understanding the Threat, Responding to the Challenge. New Haven and London: Yale University Press; 2002. p.124.

[12] European Court of Human Rights. Case-law The Republic of Ireland v. The United Kingdom [Internet]. 1978. Available from: http://lawofwar.org/Ireland_v_United_Kingdom.htm [Accessed: 2016-07-25].

[13] United Nations High Commissioner for Human Rights. Interpretation of torture in the light of the practice and jurisprudence of international bodies [Internet]. 2011. Available from: http://www.ohchr.org/EN/Issues/Torture/UNVFT/Pages/Documentation.aspx [Accessed: 2016-07-25].

[14] United Nations High Commissioner for Human Rights. Report of the Special Rapporteur on torture and other cruel, inhuman or degrading treatment of punishment E/CN/.4/2006/6. 2005, p. 13.

[15] Dershowitz A.M. Why terrorism works. Understanding the Threat, Responding to the Challenge. New Haven and London: Yale University Press; 2002. p.151.

[16] Anderson B. We not only have a right to use torture. We have a duty [Internet]. 2010. Available from: http://www.independent.co.uk/voices/commentators/bruce-anderson/bruce-anderson-we-not-only-have-a-right-to-use-torture-we-have-a-duty-1899555.html. [Accessed: 2016-07-25].

[17] Dershowitz A.M. Why terrorism works. Understanding the Threat, Responding to the Challenge. New Haven and London: Yale University Press; 2002. p. 146.

[18] Krauthammer C, The Truth about Torture, it's time to be honest about doing terrible things [Internet]. 2005. Available from: http://www.weeklystandard.com/Content/Public/Articles/000/000/006/400rhqav.asp?pg=1 [Accessed: 2016-07-25].

[19] Bagaric M, Clarke J. Not Enough Official Torture in the World? The circumstances in which torture is morally justifiable. San Francisco: University of San Francisco Law Review. 2005; vol. 39; 614: 581-1073.

[20] Levin M. The case of torture [Internet]. 1982. Available from: http://people.brandeis.edu/~teuber/torture.html [Accessed: 2016-07-25].

[21] Casebeer WD. Torture interrogation of terrorists: a theory of exceptions (with notes, cautions, and warnings). In: Shanahan T, editor. Philosophy 9/11: Thinking about the war on terrorism, Chicago: Open Court; 2005. p. 244-259.

[22] Thomas EB. Medical ethics on the battlefield: the crucible of military medical ethics. In: Thomas EB, Sparacino LR, editors. Military Medical Ethics. Washington, DC: TMM Publications; 2003. vol. 2, p. 397: 369-420.

[23] Dershowitz A.M. Why terrorism works. Understanding the Threat, Responding to the Challenge. New Haven and London: Yale University Press; 2002. p. 152.

[24] Dershowitz A.M. Why terrorism works. Understanding the Threat, Responding to the Challenge. New Haven and London: Yale University Press; 2002. p. 159.

[25] Dershowitz A.M. Why terrorism works. Understanding the Threat, Responding to the Challenge. New Haven and London: Yale University Press; 2002. p. 144.

[26] Weizman E. Il male minore. Roma: Nottetempo; 2009. p.7.

[27] Ignatief M. The lesser evil, Political ethics in an age of terror. Princeton: Princeton University Press; 2004. p.24.

[28] Ignatief M. The lesser evil, Political ethics in an age of terror. Princeton: Princeton University Press; 2004. p.8.

[29] Ignatief M. The lesser evil, Political ethics in an age of terror. Princeton: Princeton University Press; 2004. p. 42. However, we recall that the author is inclined to accept heavy forms of coercion but he does not admit any form of torture.

[30] Ignatief M. The lesser evil, Political ethics in an age of terror. Princeton: Princeton University Press; 2004. p. 9.

[31] Bagaric M, Clarke J. Not Enough Official Torture in the World? The circumstances in which torture is morally justifiable. San Francisco: University of San Francisco Law Review. 2005; vol. 39; 607: 581-1073.

[32] Lauritzen P. The Ethics of interrogation. Professional responsibility in an Age of Terror. Washington, D.C.: Georgetown University Press; 2013. p. 94-95.

[33] Bagaric M, Clarke J. Not Enough Official Torture in the World? The circumstances in which torture is morally justifiable. San Francisco: University of San Francisco Law Review. 2005; vol. 39; 607: 581-1073.

[34] Lepora C, Danis M, Wertheimer A. No exceptionalism needed to treat terrorists. The American Journal of Bioethics. 2009; 9 (10); 54: 53-54.

[35] La Torre M, Costerbosa ML. Legalizzare la tortura? Bologna: Il Mulino Saggi; 2013. p. 138.

[36] La Torre M, Costerbosa ML. Legalizzare la tortura? Bologna: Il Mulino Saggi; 2013. p. 144- 145.

[37] Bagaric M, Clarke J. Not Enough Official Torture in the World? The circumstances in which torture is morally justifiable. San Francisco: University of San Francisco Law Review. 2005; vol. 39; 615: 581-1073.

[38] La Torre M, Costerbosa ML. Legalizzare la tortura? Bologna: Il Mulino Saggi; 2013. p. 163.

[39] Arendt H. Responsibility and judgment. Trad it. Responsabilità e Giudizio. Torino: Einaudi; trad. Tarizzo D, 2004. p.31.

[40] Weizman E. Il male minore. Roma: Nottetempo; 2009. p.24.

[41] Weizman E. Il male minore. Roma: Nottetempo; 2009. p.31.

[42] Lauritzen P. The Ethics of interrogation. Professional responsibility in an Age of Terror. Washington, D.C.: Georgetown University Press; 2013. p. 103.

[43] Lauritzen P. The Ethics of interrogation. Professional responsibility in an Age of Terror. Washington, D.C.: Georgetown University Press; 2013. p. 100.

[44] Lauritzen P. The Ethics of interrogation. Professional responsibility in an Age of Terror. Washington, D.C.: Georgetown University Press; 2013. p. 101.

[45] La Torre M, Costerbosa ML. Legalizzare la tortura? Bologna: Il Mulino Saggi; 2013. p. 66.

[46] La Torre M, Costerbosa ML. Legalizzare la tortura? Bologna: Il Mulino Saggi; 2013. p. 126-127.

[47] Schmitt C. Political Theology. Four Chapter on the concept of sovereignty. Schwab G. editor and trad. Chicago: University of Chicago Press; 2005.

[48] Agamben G. Stato di eccezione. Torino: Bollati Boringhieri; 2003. p.63. Trad. en. State of exception. Chicago, London: University of Chicago Press, 2005.

[49] La Torre M, Costerbosa ML. Legalizzare la tortura? Bologna: Il Mulino Saggi; 2013. p. 64-65.

[50] Crigger B.J. Cases in Bioethics. 3rd edition Boston: Bedford/ST Martin; 1998. p. 294.

[51] Bagaric M, Clarke J. Not Enough Official Torture in the World? The circumstances in which torture is morally justifiable. San Francisco: University of San Francisco Law Review. 2005; vol. 39; 735-736: 581-1073.

[52] Allhoff F. Physician involvement in hostile interrogations. Cambridge Quarterly of Healthcare Ethics. 2006; 15; 392: 392-402.

[53] Allhoff F. Physician involvement in hostile interrogations. Cambridge Quarterly of Healthcare Ethics. 2006; 15; 395, 392-402.

[54] Gross ML. Bioethics and Armed Conflict. Massachusetts: The MIT Press; 2006.

[55] Gross ML. Bioethics and Armed Conflict. Massachusetts: The MIT Press; 2006. p. 15.

[56] Gross ML. Bioethics and Armed Conflict. Massachusetts: The MIT Press; 2006. p. 31.

[57] Gross ML. Bioethics and Armed Conflict. Massachusetts: The MIT Press; 2006. p. 63.

[58] Allhoff F. Physician involvement in hostile interrogations. Cambridge Quarterly of Healthcare Ethics. 2006; 15; 394: 392-402.

[59] Allhoff F. Physician involvement in hostile interrogations. Cambridge Quarterly of Healthcare Ethics. 2006; 15; 394: 392– 402.

[60] Lauritzen P. The Ethics of interrogation. Professional responsibility in an Age of Terror. Washington, D.C.: Georgetown University Press; 2013. p. 124.

[61] Callahan D. editor. The goals of medicine. Setting new priorities. Hastings Center Report, Special supplement, 1996. S19.

[62] Matthews R.M. Indecent medicine: in defense of the absolute prohibition against physician participation in torture, The American Journal of Bioethics. 2006; 6(3). W34-W44; W38-W39. DOI: 10.1080/15265160600686372.

[63] London L, Rubenstein SL, Baldwin-Ragaven L, Van Es A. Dual loyalty among military health professionals: human rights and ethics in times of armed conflict. Cambridge Quarterly of Healthcare Ethics, 2006; 15, 386: 381-391.

[64] Koch T. Weaponising medicine: "tutti fratelli, no more. J Med Ethics, 2006; 32, 250: 249-252. DOI: 10.1136/jme.2006.016485.

[65] Lauritzen P. The Ethics of interrogation. Professional responsibility in an Age of Terror. Washington, D.C.: Georgetown University Press; 2013. p. 99.

[66] Miles S.H. Oath Betrayed Torture, medical complicity and the war on terror. New York: Random House; 2006. p. 138.

[67] La Torre M, Costerbosa ML. Legalizzare la tortura? Bologna: Il Mulino Saggi; 2013. p. 161.

[68] Rejali D. Torture and Democracy. Princeton: Princeton University Press; 2007. p. 24.

[69] Rejali D. Torture and Democracy. Princeton: Princeton University Press; 2007. p. 455-458.

[70] Lauritzen P. The Ethics of interrogation. Professional responsibility in an Age of Terror. Washington, D.C.: Georgetown University Press; 2013. p. 87.

[71] Rodriguez CD. I Codici Deontologici dei professionisti della salute e la Guerra. Riv Diritto e le Professioni Sanitarie. 2001; 4 (3): 212-218.

[72] Lauritzen P. The Ethics of interrogation. Professional responsibility in an Age of Terror. Washington, D.C.: Georgetown University Press; 2013. p. 145.

Chapter 10

Truncated Autonomy: Neocortical Selves, Reverse Reductionism and End-of-Life Care

Ger Wackers

Additional information is available at the end of the chapter

http://dx.doi.org/10.5772/66044

Abstract

In professional guidelines for palliative sedation in end-of life care, a particular notion of conscious life experience is associated with specific cognitivist notion of frontal lobe autonomy. Drawing on Turner and Fauconnier's work in cognitive linguistics I argue in this chapter that even our most central notions like human subjectivity and autonomy are conceptual blends. This chapter explores the origins and emergence of these concepts and their entailments. It digs deep into the conceptual blending of the ontogenetic development of the individual with the phylogenetic history of life. This hyper-blend of the flesh is contrasted with the hyper-blend of an irreal, non-material deep, inner space that is co-extensive with consciousness and with the rational, operative agent constituting the human subject. The last part of the chapter explores the frictions and problematic entailments of these different hyper-blends for end-of-life care practices concerning brain death, persistent vegetative state and palliative sedation. Despite respect for a patient's autonomy being first among the principles of medical ethics, cognitivist criteria used in the assessment of a patientâ€™s decision-making competence reduce and constrain (truncate) the patient's autonomy in a variety of ways in one of the situations in life where it should matter most, in dying.

Keywords: cognitivist neocortical autonomy, conceptual blending, end-of-life care

1. Introduction

In this chapter, I will develop the analysis of a problematic that cropped up in my previous work on the role of palliative sedation in end-of-life care [1, 2]. Namely, the ways in which professional guidelines constrain the autonomy of patients in one of the situations where it matters most, that is in the process of dying. End-of-life care is an important object of governance. In the governance of end-of-life care, there are two 'objects of concern'. One is to protect the autonomy and integrity of the dying patient. The other is to maintain public trust in the institution of modern medicine and professional health care. The two are related of course. To maintain public trust in professional health care, constraints are imposed on what assistance-in-dying you can ask for, what will be offered to you or granted when you ask for it. Respect for the patient's autonomy ranks highest among the principles of medical ethics [3]. Yet, when it comes to dying, the patient's right to self-determination is constrained by professional judgment and the health-care professional's right to conscience. A fine line runs between allowing to die and hastening the patient's death. The boundary between the two is diligently monitored and protected. It is a borderland between the prevention and prosecution of illegal, criminal killing and compassionate care, which does not stop when lives can no longer be saved or prolonged. I was struck by the central importance of a notion of 'conscious life experience' in the arguments *against* a more generous availability of palliative sedation in end-of-life care and by a related operationalization of informed consent in terms of higher, neocortical functions of information processing and decision-making. The respect for *a particular form* of *neocortical autonomy* and the almost unbounded respect for the sanctity of life form an unbreakable couple, culminating in the paramount centrality of consciousness in professional guidelines for palliative sedation. It is in a sense peculiar that these principles should stand so strong in end-of-life situations when disease processes erode the capacity for integrated functioning of the brain and body and when death is near. Is it because we cannot let go when it is time?

2. Conceptual blending

To make an inroad into this problematic I will start from the pre-position that all judgments about the world, about human nature and core human values are the products of *cognitive* processes embedded in the bodies, brains and activities of individuals and the interactions among and practices of members of collectives. I will pursue an analysis that draws on recent work in cognitive linguistics. More specifically, I will draw on theories of conceptual blending or conceptual integration in meaning construction as advanced by Fauconnier [4–6] and Turner [6–8]. Conceptual blending theorists do not distinguish between the blending that occurs on the fly in everyday meaning construction through language and dialog [9] and scientific theories and philosophical models [6, 7]. All forms of scientific thought and reasoning can be subsumed under the banner of human cognition, to which perspectives, theories and concepts of embodied and distributed cognitive science may be fruitfully applied. Following Turner we could argue that every theoretical concept is a hyperblend, including the notions that interests us most, namely the notions of human agency and autonomy, *recognizing* them—in the polysemic meanings of both *knowing* and *acknowledging*—as products of contingent, fundamentally biological and social *cognitive* processes.

Let me first explain conceptual blending by example. We usually think of how old we are in terms of chronological age (CA). Measured in years and months, CA measures the time elapsed since we were born. Always living in the present, CA is a point moving along the arrow of time, neither turning back on itself nor speeding up into the future. Biologists of aging have

http://dx.doi.org/10.5772/66044

introduced the concept of biological age (BA) and search for ways to measure it [10]. In a normalized population, they have measured different sets of features of the body (biomarkers) and how they change with age. Based on these population data, they have construed an alternative aging scale to which an individual's values can be compared. Derived mathematical algorithms are used to calculate the age with which these individual values correlate. In forensic science, such scales based on, for example, development features of the skeleton and dentition are used to estimate the chronological age of a person or body for which that information is missing or disputed. In public health practices, this computed BA is compared (blended) with your chronological age, as a result of which you may be 28-year-old with the body of a 65-year-old. The average life expectancy that has been calculated (blended and compressed) for the men or women who were born around the same time as you were, has turned your lifespan into a limited amount of time that you have been using up at an accelerated speed; destined for a *premature death*? [11] The good news is that BA, contrary to CA, is reversible. The *modifiable risk factors* that have been entered into the etiological model explaining the discrepancy between an individual's CA and BA are within the range of the individual to do something about. By implication, personal health risk management is the individual's responsibility.

Conceptual integration can be schematically presented by the way of a minimal network that comprises at least three mental spaces: at least two input mental spaces and the blending space [8]. The input spaces selectively contribute or project structure and elements to the blended space in which these are integrated. Structure and elements that occupy analogous positions in the two input spaces, which in other words map between domains, may be compressed into identity and human scale. The individual in the CA blend is identical to the individual in the CA blend. However, there may also be vital disanalogies in the blended space: the discrepancy between CA and BA. Following its initial integration structure and elements in the blend can be elaborated, as a result of which new structure and meaning emerges that was not initially available from the input spaces: *premature* death, the idea that you might die ahead of your time, earlier than necessary. In this example, the blended space presents a trajectory of life that is counterfactual compared to the one on which you now are going: a trajectory in which it is possible for you as an individual to reach the average life expectancy of your cohort. This trajectory will remain counterfactual unless you take care and take responsibility for your own health. The point is not that the blend is a 'possible world' or a true representation of the world, but that the blend suggests alternative ways of engaging with the world, and whose primary responsibility that is.

Conceptual integration performs work on its previously entrenched products. New blends are made out of inputs that already are blends. That is, in its advance forms, conceptual integration makes *blends of blends*. Turner calls these "blends of blends"... hyper-blends. "A hyper-blend is a blend that has at least one of its input mental spaces something that is already a blend. Hyper-blends occur routinely" [8].

Models of cause-effect relationships are an important type of blend that enters into hyper-blends that are more complex. In human scale situations, we may observe events in close spatiotemporal proximity and construe one as the cause of the other. In many situations, however, we only have records of what we take to be effects and try to learn about causes (antecedents) by making inferences from these preserved results (consequences). However, we are particular about what we allow to take the role of a cause. We usually do not allow events that occur after what we take to be an effect to act as a cause. However, we do allow the passage of time to be the cause of death when we say that a person 'died of old age'. We allow features, actions and events that did not happen (counterfactuals) to play the role of contributing causes when we say that an omission of or failure to install a preventative measure causes a disease. We allow private and immaterial human intentions to have effects in the material world or infer unobservable (and difficult to prove) private intentions from recorded effects in the material world. Through blending, we routinely amalgamate the physical and mental, material and immaterial, factual and counterfactual, and is and ought. We blend to create meaning, shape and understand the world we live in.

3. Living in the blend

A theory of conceptual blending is, however, not just a handy tool for conceptual analysis. As a 'space of representation' [12, 13], the blended space envelops, in addition to concepts and logical structure, also data that are produced under the blend's specific epistemic conditions or selectively projected or appropriated into the blend to provide empirical support or foundation. "Of course", Fauconnier [5] writes, "observation and theory are part of the same overall package; a 'phenomenon' requires a theory, even if it is a folk theory, in order to be observed at all. There is no absolute, direct theory-independent observational interpretation of the 'facts'. As a science evolves, there is simultaneous, parallel evolution of the observational procedures and interpretations, and of the explanatory theory itself". Furthermore, conceptual structure, framing logic and produced data are blended with ontoepistemological positions. When blended with a realist or essentialist epistemology, theoretical concepts that were introduced to help order and make sense of disparate data or events reify into real-world processes. Inferred causes are assumed to exist in the world, past or present, and exert their forming power on recordable effects. Human rights and values are assumed to derive from the essential nature of human beings and cannot be alienated from or denied beings who belong to that category. Theoretical notions, produced data and epistemological position thus form, when mutually supportive, a robust package, a *mode of ordering* to use Mol and Law's term [14]. As temporarily stabilized formations, 'modes of ordering', or hyperblends, do not only construe but also represent worlds. Erasing the faint line between epistemology and ontology, one could say that it (the hyperblend) is the world. For scientists, or philosophers, or practitioners with a 'high inclusion in the frame' [15], the set of claims that make up an advanced hyperblend lies too close to the core of their deeply assimilated and largely unconscious beliefs to be challenged, or even overtly recognized as something potentially disputable. They live in the blend. There is recursivity here. Collectively and over time, we construe worlds – through cognitive processes of hyperblending concepts, data and epistemology – and then live in them. However, 'from time to time', Rheinberger [13] writes, "new forms emerge that have something significant about them, something that catalyzes previously present actors,

http://dx.doi.org/10.5772/66044

things, institutions into a new mode of existence, a new assemblage, an assemblage that not only puts things in a different light, but makes them work in a different manner".

The power a particular hyperblend can hold over people, fueling its own protection and the rejection of alternatives, should not be underestimated. Apparently, this is the case when deeply engrained beliefs about our own human nature are challenged; about what sets us apart from animals; about what is our essence at the end, when our existence as an individual human being is imminent and about what remains after our body has eroded and disintegrated. But neither should the importance be underestimated of recognizing the contingency, multiplicity, diversity and simultaneous coexistence of different modes of ordering life and the world, and how they relate to and interfere with each other. "For", Mol and Law [14] argue, "the various modes of ordering, logics, styles, practices, and the realities they perform do not exist in isolation from one another ... They are not islands unto themselves, closed cultures, self-contained paradigms, or bubbles. [T]hey interfere with one another and reveal ... partial connections". 'Often', Mol and Law [14] continue, "it is not so much a matter of living in a single mode of ordering or of 'choosing' between them. Rather it is that we find ourselves at places where these modes join together. Somewhere in the interferences something crucial happens ... complexity is created, emerging where various modes of ordering (styles, logics) come together and add up comfortably or in tension, or both". The history of the sciences, philosophies and ethics is a history of diversification of modes of ordering, not of the successive replacement of one by the other. It is a history of contended claims about the appropriateness of hyperblends and their entailments.

Blending helps us to compress to human scale and handle, to use Turner's favorite phrase, "vast ideas that span great ranges of time, space, causation and agency, ideas that are not at all restricted to the local scene" [8]. The blended space does not replace the spaces from which it received inputs. In a methodological sense, this implies that the input spaces are accessible and amenable for analysis *through* the blend. However, there is also a diachronic, historical dimension to hyperblends. Powerful hyperblends have evolved over time and can be traced in the history of the sciences and of ideas. To trace the evolution of modern notions of human subjectivity and autonomy we must engage with the history of ideas about the relationship between the development of an individual and the history of life.

4. Blending development of an individual with the history of life

A fault line roughly separates the hyperblends that concern us here, comprising key notions about human nature, consciousness and autonomy. On the one hand we find blends that feature a deity, or other supranatural being, as the creator and giver of life (creationism, intelligent design). On the other hand, we have views of life advanced by sciences that do not allow as causes, entities or processes that are not part of the natural world [16]. Neither do they allow outcomes that emerge temporally after their antecedents (no teleological explanations allowed). I recognize the importance of a spiritual dimension in palliative care and the strength many people derive in dying from their belief in a higher being and an afterlife. However, I am primarily interested in the power and centrality of an idea of conscious life experience that overrules a person's own direction over the process of dying in a secular, nonreligious context. Therefore, I will leave the former aside and pursue the latter. This hyperblend or view of human life is grafted on a compression into unity of the development of an individual organism and the history of life on earth. On a human scale, we experience individuality as the membrane or skin bound discreteness and uniqueness of organisms. In human reproduction, animal husbandry and plant breeding we experience the birth, growth and decline of individual organisms on a time scale for which the duration of a human life provides the standard. Endeavors to understand the nature of the history of life on earth, stretching back into deep time measured on a geological timescale and utterly inaccessible for present day biologists, have not only received inputs from the study of fossil records [17], but also from the comparative anatomy of now-living species and from embryology. Especially the latter, the idea the one can learn about the history of life from the comparative study of the morphology of embryological development across species has contributed strongly to the blending of ontogeny and phylogeny [18]. Recognizing that all species that rely on sexual reproduction for their procreation pass through a single-cell stage, this fertilized egg maps as the origin of an individual organism onto the origin of life. The 'unfolding' of the individual organism through fetal growth, birth and postnatal maturation maps onto the 'unfolding' (evolution) of life from its humble unicellular origins through higher levels of complexity and perfection in the arrival of man. The themes of differentiation, specialization and maturation of cell lineages and tissues in developing individual embryos map onto the diversification of life forms through speciation and the emergence of excellence in the branching tree of life.

This blend has not only inspired an 'iconography of an expectation' [19], which is the more-than-familiar present day cartoons that represent the evolution of man as a linear march of progress but also inspired the nineteenth century biologist Ernst Haeckel's biogenetic law, better known as the notion of 'recapitulation': the idea that ontogeny recapitulates phylogeny and that embryos during their individual growth and development pass through the adult forms of its ancestors. In Haeckel's view, evolution occurred through the heritable addition of new features to the adult form of ancestors [18]. In other words, "we climb our evolutionary tree in the womb" [20]. Haeckel published a famous plate showing a series of embryos, each aborted in its own development, in a grid along two dimensions: an ontogenetic series comparing stages of an individual's development and an evolutionary series comparing different species. In Haeckel's 'space of representation', in his blend, individual organisms deliberately selected and ordered in a two-dimensional series came to express as new and emergent features in the blend namely development and evolution, history and progress. "Development", Hopwood [20] argues, "is thus not simply the process embryologists study; it is also an effect they have labored to produce". Haeckel's plates gave embryos presented as developmental series a public profile [20]. The idea of ontogeny recapitulating phylogeny endured far into the twentieth century and exerted its influence long after it had fallen from grace and had lost credibility among professional evolutionary biologists [18].

The hyperblend of individual development with the history of life, characterized by directed evolution towards an expected arrival of man at the pinnacle, has been challenged by alter-

http://dx.doi.org/10.5772/66044

native blends. Darwin displaced the idea that the 'unfolding' of an intrinsic essence was the primary, causal force driving evolution in the history of life. He replaced it with the idea of *natural* selection, itself a blend with input from *artificial* selection as experienced on a human scale in animal husbandry and plant breeding, Natural selection works on undirected, isometric variation (equal likelihood in all directions) in a population. Gould [21] points out that for Darwin natural selection was not just an external negative force weeding out the misfits, but that it was a creative, positive force in and cause of evolution working gradually over vast expanses of time. For Darwin, natural selection worked on the level of organisms struggling for survival. In the twentieth century, Stephen Jay Gould, with Niles Eldredge, worked to recast, through their theory of punctuated equilibrium, the history of life as "a story of massive removal [large extinctions] followed by differentiation within a few surviving stocks" [19]. Gould also worked to expand Darwinian evolutionary theory into a hierarchical theory, allowing natural selection to operate, above the organismal level, on the level of species. In a section called 'The Grand Analogy' in *The Structure of Evolutionary Theory* [21], Gould engages in an extensive and explicit blending operation. Gould maps individual organisms as the units of selection in microevolution to species as the units of selection in macroevolution. He maps the birth and death of individual organisms to the origin and extinction of species. He maps the timescale of the life of an individual to the scale of geological time; the parent–child relationship of organisms to parental and daughter species, the transmission of features from parental generations to offspring to the transmission of formative properties between ancestral and descendant species. Macroevolutionary trends "could be conceptualized as a result of higher order selection upon a pool of speciational events that might occur at random with respect to the direction of the trend. In such a case, the role of species in a trend would become directly comparable with the classical status of organisms as units of change within a population under natural selection" [21]. In this blend, the history of life is not a linear march of progress towards an expected outcome (the arrival of man), but a story that blends ontogeny and phylogeny in a different way.

5. Heterochrony in human development and the brain

Thus far, ontogeny seems to have provided a conceptual structure to phylogeny, the human scale experiences and observations of discrete organisms serving as the source domain of metaphorical projection of structure to the target domain: the utterly inaccessible domain of the history of life. However, Turner [8] emphasizes how structure from the blended space can be projected back to the input spaces. Despite disagreements and controversies among evolutionary biologists about the precise nature, relative frequency and importance of formative, causal processes, the conceptual blends achieved have recast the human species as a species among other species with whom it shares common ancestors. Humans share with other vertebrates the same fundamental 'building plan'. Humans share the basic structure of their brains with other mammals. Our brains are mammalian brains.

It is in our 'expanded' brains, and the 'higher' cognitive functions that they afford, that many scientists and philosophers locate the features through which the human species bootstrapped itself out of the world of brute animals. Yes, we belong to the animal kingdom, we share ancestors with our companion species, but we also stand above and at a distance from them. Exactly what those features are and what brought them about is a matter of contention.

Today, there seems to be little support among evolutionary biologists for the idea that the expansion of the human brain can be accounted for by the accumulation of mutations in structural genes, that is, in genes that code for proteins. The number of structural genes in the human genome does not far exceed that in other primates. In other words, the differences in genotype are insufficient to account for the marked differences in phenotype. However, in recent years, more emphasis has been put on the regulation of gene expression – both on genes regulating the expression of other genes and molecular non-DNA processes involving histones – the proteins around which strands of DNA are winded to form chromosomes. Interestingly, echoing this shift in emphasis, Gould [18] argues for the relative frequency and factuality (sic!) of heritable changes in the rate of ontogenetic development to account for macroevolutionary trends on the level of species. These changes in the rate of development may be different for different parts of the body and affect developmental processes like teething and the onset of sexual maturation differently.

This 'heterochrony' in ontogenetic development is, when we think about it, a familiar feature. We know from embryology that the cranial parts of the embryo, encompassing the brain, develop and grow faster than its caudal parts. We can observe this feature in common ultrasound images of first-trimester pregnancies. Even at birth the baby's 'hind-limbs' are relatively undeveloped compared to its head, making the newborn infant dependent on maternal and parental care for a long time after birth. The caudal parts of the body catch up during childhood, developing and growing into long and strong legs before the onset of sexual maturation blocks further growth. Contrary to what one might expect, the evolution of the human species and its expanded brain is not the result of a speeding up but rather of a slowing down of developmental processes: *retardation*. This change in the rate of development delays the onset of processes that constrain further growth, allowing for longer growth periods. The slower rate of development and maturation also correlates with a longer lifespan. This relationship seems to hold for the multicellular organism, as well as for the differentiated cell types within the organisms. Slowly maturing nerve cells are the longest living cells of the human body. Fast developing red blood cells, which even discard their nucleus before entering the peripheral blood stream, are among the cells with the shortest lifespan. In humans, newborn infants maintain a fetal growth rate after birth for about a year, much longer than other newborn primates do. Compared with other species humans enjoy the longest juvenile periods before sexual maturation sets in Ref. [18]. The reduced rate of development allows the brain to grow longer and extend its patterns of cell proliferation in time. As a result, humans achieve a markedly higher degree of encephalization. The delay in development also delays the loss of plasticity that is correlated with maturation.

Compared with other species, humans retain for a longer period of time the capacity to form new neuronal connections and to modify existing ones. Jacobson [22] argues that "this reduction [of plasticity] occurs at different times in different classes of neurons, so that those which are generated late in ontogeny and those which mature slowly have the greatest degree

http://dx.doi.org/10.5772/66044

of modifiability in the mature animal". The long drawn-out growth period, the prolongation of childhood and the retainment into adulthood of neural plasticity under conditions of parental care and instruction, supported by the faculties of memory and language, make 'human' a learning rather than an instinctive animal [18]. It is the embodied activity of the living organism that shapes the pattern of synaptic connections and their firing properties. Hebb [23] argued that "What fires together wires together!"; the principle of what has become known as 'Hebbian learning' that underlies the formation and reformation of neural cell assemblies. Explicitly using Burnet's clonal selection theory of the immune response as an input to the blend, Edelman [24–26] developed a theory of neural group selection to account for the ways in which the pattern of neural connections is shaped through 'differential amplification' in recursive reentrant systems by the activities and experience of the living organism, not by prior programming.

6. Hyperblend of the flesh: self-building brains in evolutionary history and individuals

The ontogeny/phylogeny blend is extended into the morphology of the brain and further blended with inputs from new domains, from the study of consciousness and cognition. This blending process is characterized by a progressive reduction or truncation of the notion of cognition.

Embodied cognition biologists argue that cognition is a fundamental feature of biological life [27–29]. Cognition is coextensive with the recursive sensorimotor loops of the embodied activity of living organisms. In this blend: life = embodied action = cognition. Cognition is a function that emerges with the formation of a living organism. Cognition is not a feature that is confined to the emergence in human brains of self-consciousness. The basic story goes like this. With the formation of a membrane, that envelops an intracellular space and separates it from a now-external environment, any living organism must maintain its functional integrity and organization (homeostasis) in interaction with changes in its environment. This is a recursive relationship. Devices in the membranes (receptors) allow the organism to sense changes in its environment. Internal, or external but membrane bound motor devices (cilia, flagella) allow it to respond to these changes. The repeated, recursive cycles of action and perception constitute an intentional arc. In other words, intentionality conceived as an organism's orientedness towards its environment is a property emergent from the formation of a membrane. Multicellular organisms developed specialized cells and structures for these sensory and motor functions. Only organisms that live free and mobile in their environment have developed a nervous system. A central nervous system and brain developed from the gathering at one side of the body (encephalization) of specialized neurons monitoring and regulating changing conditions inside the body (giving rise to emotions), in close association with sense organs and neurons monitoring changes in the environment. At some point in this evolutionary history, consciousness emerges, at first in prototypical forms that also other animals may have, then blooming into a fully-fledged autobiographical human self-consciousness.

Exactly how body/brains make minds and selves, let alone why, is still a matter of contention. There is dispute with regard to the how and when they emerged in the history of life. There are many theories trying to grasp how brains make mind and usually a single self in the individual body. Nevertheless, the tentative answers to these elusive questions, including the hierarchical relationship that is assumed to exist between humans and animals, between reason and emotions, have been blended with the morphology of the brain. In ascending order, the brainstem encompasses the neural centers that regulate respiration, circulation and temperature. These are the 'vegetative' functions that have to do with the internal conditions of the body. The midbrain is associated with basic positive emotions like care and joy, and with negative emotions like fear and anger [30]. Although brainstem and midbrain emerge during organogenesis in the first trimester of pregnancy, together with the cerebral hemispheres, in a phylogenetic sense they are considered to be the oldest parts of the brain that we share with animals. The two hemispheres with their outer rim, the cortex, are considered to be phylogenetically younger and newer, hence *neocortex*. In addition to visual and motor cortices, that must be assumed we also share with other species, it is here in the neocortex we locate the functions that make us distinctly human: the areas for language perception and language generation, for abstract thinking and reasoning, for imagining alternative counterfactual scenarios and evaluating possible outcomes, yes, for conceptual blending in all its various forms. Especially the frontal lobes, above our eye sockets, stand out as the locus of the most advanced, executive higher brain functions. These functions are not only higher in the sense of more advanced then vegetative functions or emotional functions. They are also phylogenetically higher because they make us distinctly human.

The macroanatomical structure of the individual brain has by evolutionary neuroscientists been overlain (blended) with the emergence through evolutionary time of what Damasio calls a protoself or protoconsciousness, a core self-and-consciousness and the fully-fledged human *autobiographical* selfconsciousness [31–34]. Autobiographical consciousness is the defining feature of what makes us living human beings. It is called autobiographical because it locates the current self in a temporal continuum of a lived past (memory functions) and an anticipated but open future (functions of planning and evaluation of alternative courses of action) [34]. All this miraculous mental and conscious activity, including the sense of unity and continuity over time of an autonomous Self is, in one way or the other, the result of ongoing dynamic biological processes in here-and-now brains in continuous interaction with the body in which it is embedded and of the body's physical and social environment. In this hyper-blend of the flesh, the Self is 'a perpetually recreated neurobiological state' [31].

7. Hyperblend of the immaterial, autonomous, human subject

The modern notion of the autonomous human subject has received inputs from several distinct domains. Rose [35] argues that "the sense of ourselves as "psychological" individuals developed across the twentieth century … we came to understand and act upon ourselves as beings inhabited by a deep internal space shaped by biography and experience, the source of our individuality and the locus of our discontents". We locate psychology's deep internal space

http://dx.doi.org/10.5772/66044

intuitively inside our skull – above and behind your eye sockets, the space occupied by your frontal lobes? We know from elementary human anatomy that it is the brain that occupies the skull. There is no empty space there. Yet, this deep internal space is the seat of the self-governing subject. This concept underlies the important humanist notion of personhood and human agency. This line of thinking has a long and august pedigree that can be traced back to John Locke and his contemporaries. In his *Essay on Human Understanding,* Locke defines a 'person' as 'a thinking, intelligent Being', that has reason and reflection, and can consider it self as it self, the same thinking thing in different times and places" [36]. We conceive of the deep, internal, private space of the individual mind as being coextensive with consciousness and with the operative agent doing the intending, willing, emoting, conceptualizing and associating of concepts to language that we associate with thinking. It is the thinking that we do in this internal space, and of which we are consciously aware, that we have come to define – in the sense of *delineate* – as human cognition. In the same vain, it is the intentions of which we become consciously aware that we have come to define as human intentionality.

Input to the hyper-blend of the immaterial human subject has also come from a research program in cognitive science that emerged in the 1940s. This program received inputs from mathematical logic, general systems theory, Shannon's statistical information theory of signals and communication channels, and not in the least, from the invention of information-processing machines such as digital computers. An important entailment of this program was that cognition, conceptualized as information processing, storage and retrieval, was *platform independent*: it had been instantiated by Nature in the biological tissues of the brain, but it could also be instantiated by engineers in artificial silicon-based devices. Like the operation of software, cognition was independent of the hard ware on which it ran. Again, conceptual structure and logic was projected back from the blended space to the input space of human cognition. This blend drove a wedge between human cognition and the human body/brain. It displaced the body and disembodied cognition. It also displaced emotions as a form of cognition, that is, as the human body's principal biological valuation system. However, it embraced technologies that now became 'intelligent' and 'smart'. Despite the criticism that has been leveled by embodied cognition theorists [27–29] against this 'cognitivist' program, it has been extremely influential in cognitive science and theories of learning.

As a third input to the blend, Woods [36] argues that the now-commonplace notion of the autonomous self has evolved out of a backdrop of rebellion against traditional sources of authority such as the church and state. 'The contemporary concept of the 'self' and 'person', Woods argues, 'is bound up with a politics of non-interference into the personal life of the individual…' [36]. Autonomy as a right to non-interference is firmly embedded in the now-stronger than ever requirement of informed consent in medical practice and research. In informed consent procedures, cognition is operationalized in 'cognitivist' terms as information processing. Based on valid information provided by health-care professionals the patient weighs, evaluates, judges and makes a decision of consent or approval concerning the medical procedures that are proposed being done to her. The information provided being understood is a prerequisite for the consent to be valid. In other words, the validity of the consent is a function of the information provided. I have called the notion of autonomy embedded in and performed through these informed consent procedures for *neocortical autonomy* [2]. Perhaps *frontal lobe autonomy* is a better term. Both terms express the *blendedness* of this highly cognitivist concept of autonomy with the parts of the brains anatomy on which it is projected.

8. The *irreality* of the human Self: partial relations, frictions and problematic entailments

Thinking about psychology's deep internal space, in which we locate a cognitivist Self that is the source of motives, intentions and the ability to choose to do what is right – which defines us as moral agents -, I have struggled to find a term to characterize a space that is real in one blend but does not exist in another. How can the human subject, devoid of any form of materialness, exist in human anatomy with its hierarchical spatialization of the brain that is grafted on ideas about the species' phylogenetic history? For want of a better word, I have called it an *irreal* space to indicate that it is the result on an imaginative act of conceptual blending [2]. The modern notion of the human autonomous subject (Self) may also be characterized as *irreal*: to express the ambiguousness in two senses in which we think about ourselves: (1) the sense that we are coextensive with our body, with the entailment that the 'I' or 'Self' ends with the disintegration of the body/brain, and (2) the sense that there is some-thing more, an immaterial entity (soul?) that we perhaps hope will live on after death. I do not want to define the two blends as two essentially distinct and separate ontological domains, Descartes' *res extensa* and *res cogitans*. I want to stress the complex ways in which they are related. As Mol and Law put it, 'If there are different modes of ordering that coexist, what is reduced or effaced in one may be crucial in another so that the question no longer is, Do we simplify or do we accept complexity? It becomes instead a matter of determining which simplification or simplifications we will attend to and create and, as we do this, of attending to what they foreground and draw our attention to, as well as what they relegate to the background' [14].

In everyday life, we routinely, and apparently without too much trouble, amalgamate the immaterial mind with the flesh inside our skull. The lack of trouble, I suggest, is a result of our ability to move between and inhabit them both, depending on the discursive requirements of the situation. Alternating between them helps us to avoid the complexities associated with the places where the different modes of ordering (hyperblends) join together and interfere. However, to use another of Turner's [8] favorite expressions, we are not fooled. We know that consciousness is dependent on the functional integrity of the biological processes of the brain; that lack of oxygen (syncope; ventricular fibrillation), or glucose (diabetic hypoglycemia and coma), the loss of coordination of neural firing (epileptic insult) or the inhibition of certain neurotransmitters (anesthesia), very quickly makes us loose consciousness. The reversibility of these conditions, with the phenomenological reappearance of a sense of Self, suggests *continuity* of the Self even during the period of unconsciousness: an effect that we labor to produce. We also know that psychoactive drugs can provoke hallucinations or modify the quality of our mental state: an effect that we, when positive, may seek or use therapeutically in an attempt to treat or rebalance mental disturbances. However, the frictions between the

http://dx.doi.org/10.5772/66044

two hyper-blends become inevitable in end-of-life care situations, when the 'brilliant fire of consciousness' [24] is in danger of being extinguished permanently and irreversible.

In the 1950s, when the new resuscitation techniques of external closed-chest compressions in combination with mouth-to-mouth ventilation left the hospital and were taught to lay people, the greater sensitivity of the neocortex to lack of oxygen – compared to midbrain and brain stem structures – produced a new *form*. In some cases resuscitation efforts did succeed in restoring spontaneous heart action and respiration, the regulatory centers of which are located in the brain stem. However, in the same patients the restoration to *animated* conscious life failed. These patients survived an episode of cardiac arrest, but they remained severely brain damaged: dead brains in bodies with hearts again beating. French physicians coined the term *coma dépassé* to describe the state. Today the condition is known as *persistent vegetative state* (PVS). With the widespread use of external cardiac massage and mouth-to-mouth breathing, supplemented with closed-chest defibrillation and supportive drug therapy, their numbers increased during the 1960s and the 1970s. Their undeniable presence triggered a reformulation of the concept, criteria and diagnostics of human death in terms of *brain death* [37, 38]. Definitions of *neocortical death* have been proposed, attributing central importance to the irreversible loss of *higher brain functions* [39, 40]. For these functions, a viable, phylogenetically younger, cerebral neocortex was identified as a necessary, albeit not sufficient *conditio sine qua non*. Proponents of a neocortical definition of death took great care to emphasize that the work of redefining death served to clarify the *real o ntological status* (sic!) of severely brain damaged patients in intensive care arrangements. The distinction between the body that has developed or retained the capacity for higher brain functions - the body that *is* an embodied person -, and the body that has irreversibly lost this capacity - the body that *is* a biological remnant after the person has died-, is not, so they argued, a value-based or moral divide. Yet, neocortical definitions of death have not been translated and embedded into law.

As the question of the status of severely brain damaged patients in intensive care units became more urgent as the need for good quality donor organs increased, most countries settled on a more conservative *whole brain definition of death* that requires that all nerve cells within the skull must be shown to be irreversibly damaged. Only after demonstrating that there has been no blood flow through the brain for 20 min or more can the patient be pronounced dead and organs extirpated. Practices preceded their legal regulation. In the Netherlands, a provisional legal basis for the extirpation of healthy organs for transplantation purposes was found in the *Law on the Disposal of the Dead*. As an input to the new emerging hyperblend of brain death, it was the analogy between autopsy and partial dissection on one hand, and the extirpation of health organs for transplantation purposes on the other, which provided in 1971 the initial legal justification for the latter. The Dutch legislator spent 20 years, and a series of Health Council reports, negotiating public confidence in the new brain-related definition of death before eventually formalizing the new hyperblend in the 1991 bill on the Donation of Organs [38].

Today, PVS-patients still have the ontological status of being alive and deserving of the protection of the law, despite the absence of consciousness. Whereas a stable new hyperblend, or mode of ordering, has emerged around potential heart-beating organ donors with dead brains, matters are quite different for people still in possession of a fully fledged autobiographical consciousness who wish to maintain control and a certain degree of self-direction over the time and manner of their death. With a few exceptions, physician-assisted suicide and euthanasia are illegal in most countries. Even national and professional guidelines that aim to regulate the use of palliative sedation for terminal, dying patients are quite restrictive. Criteria for this pharmacological reduction of consciousness in dying patients include intense and sustained suffering from physical symptoms that have been shown to be refractory to ordinary treatment. Arguing on the one hand that palliative sedation should be considered to belong to the repertoire of ordinary and legal medical treatments, these guidelines clearly mark, through restrictive safeguards, their distance from illegal practices of assisted suicide and euthanasia [1, 2].

Despite the status of respect for a patient's autonomy as the first among the four principles of medical ethics [3], in texts of law, professional guidelines and codes of ethics, it is the much more restrictive cognitivist, frontal lobe notion of the human subject that has been inscribed. One problematic entailment of this is that it seems as if patients, in the face of death, cannot be trusted to know their own mind. The *competence* to make end-of-life decisions has been formally delegated and assigned to physicians, reducing the patient's autonomy into a right to be heard. Respect for the patient's autonomy has been made conditional upon health-care professionals assessment of a patient's cognitive capacity to give informed consent, applying criteria derived from the same truncated cognitivist frontal lobe notion of human subjectivity and personhood. In practice, critically ill and dying patients in health-care settings are often dependent on the generosity of physicians and other health-care professionals. That is, they depend on health-care professionals' understanding of legal regulations and guidelines in combination with their own understanding of the *discretionary space* and leeway that their *professional autonomy* grants them.

9. Conclusion: autonomy truncated three ways

There is no lack of modern states' desire to govern practices of end-of-life care. There is no lack of intervention in and interference with the hardest thing we all must do, dying. This is also the hardest but perhaps most important thing we ever have to do for our partners in life or for parents, that is to assist them in dying [1]. Although first among the principles of medical ethics, the modern notion of the autonomous subject is truncated in at least three different ways. First, exercising autonomy through the direction of the manner of one's death is constrained by the right to conscientious objection of health-care professionals. This right is rooted in the same principle of autonomy as the patient's right to self-determination and noninterference that is inscribed in the international declaration of human rights. As a patient "You have and absolute autonomous right to determine what happens to you … Unless you don't!" [41].

Second, this style of thought reduces the human person to the cognitive capacities and higher brain functions of the bark (neocortex) of the frontal lobes. It ignores the massive recursive

http://dx.doi.org/10.5772/66044

connections and interactions throughout life between the neocortex and the deeper, midline structures of the emotional brain and the body that are being gradually elucidated by neuroscience [30, 32]. Natural sciences are often accused of reducing the whole person to molecular-genetic and neurophysiological events in the body and brain. In this cognitive, neocortical version of autonomy and personhood, we may recognize a kind of reversed reductionism. This is not to say that neurosciences shall have the last word, but it may be important, as Rose and Abi-Rached [42] argue, to critically explore the implications of the new sciences of the brain for the human and social sciences.

Third, cognitivist notions of an atomistic, modern autonomous subject displace relations and interactions with other beings as processes that are constitutive of human agency. There is no lack of recognition of the importance of social relations and the ability to communicate, even to the extent that this is the capability that makes us truly human. However, the atomistic, individualized person and human subject located in psychology's deep internal but irreal space gains priority. Social relations and communication come in the second place. The neocortical Self overrules any other conception of humanness. As the governor that controls the city, its stronghold and inner citadel (conscious life experience) must be protected at all costs, even – or perhaps especially – in the face of death. Simultaneously and recursively, by protecting this idea of the *autonomous subject in the patient,* health-care professionals and the law makers who hold the medical professions in high regard maintain the ethical high-ground of moral accountability.

However, we may invert these arguments and offer an alternative hyperblend for consideration. We can argue that human autonomy and independence is a temporarily stabilized result or outcome of a network of relations of dependence and recursive interactions. It is not an essentialist feature of the individual. When that stability starts to unravel due to critical illness and imminent death, the people closest to you in your network should be the ones prepared to accept some of the burden of care as the network reconfigures. We may reclaim cognition from the cognitivists and appreciate emotions again as our principal embodied value system. Because we have mammalian bodies and brains, we can explore the role of physical intimacy and touch as an important mode of caring for the dying, also for those that are no longer conscious. Thinking along these alternative lines would give access to an alternative set of practices around dying and end-of-life care, redefining the roles of the dying patients, next of kin, health-care professionals and treatment options that should be generously available.

Author details

Ger Wackers

Address all correspondence to: ger.wackers@uit.no

Department of Health and Care Sciences, UiT The Arctic University of Norway, Tromsø, Norway

References

[1] Wackers, G. Recursive health care structures and choice in the manner of our dying: an autoethnographic account. Qualitative Health Research. 2015: p. 1–14; DOI: 10.1177/1049732315576498.

[2] Wackers, G. Into the wild country: epistemic issues in professional guidelines for palliative sedation in end-of-life care. Evidence & Policy. 2015; 11, 3: p. 331–349.

[3] Beauchamps, TL and Childress, JF. Principles of Biomedical Ethics. 7th edition, Oxford: Oxford University Press; 2012.

[4] Fauconnier, G. Mental Spaces. Aspects of Meaning Construction in Natural Language. Cambridge: Cambridge University Press; 1994.

[5] Fauconnier, G. Mappings in Thought and Language. Cambridge, UK: Cambridge University Press; 1997.

[6] Fauconnier, G and Turner, M. The Way We Think. Conceptual Blending and the Mind's Hidden Complexities. New York: Basic Books; 2002.

[7] Turner, M. Cognitive Dimensions of Social Science. The Way We Think About Politics, Economics, Law and Society. Oxford: Oxford University Press; 2001.

[8] Turner, M. The Origin of Ideas. Blending, Creativity, and the Human Spark. Oxford: Oxford University Press; 2014.

[9] Coulson, S. Semantic Leaps. Frame-Shifting and Conceptual Blending in Meaning Construction. Cambridge: Cambridge University Press; 2001.

[10] Jackson, SHD; Weale, MR; Weale, RA. Biological age – what is it and can it be measured? Archives of Gerontology and Geriatrics. 2003; 36, 2: p. 103–115.

[11] Levine, ME. Modeling the rate of senescence: can estimated biological age predict mortality more accurately than chronological age? The Journal of Gerontology: Series A. 2013; 68, 6: p. 667–674.

[12] Rheinberger, H-J. Towards a History of Epistemic Things. Synthesizing Proteins in the Test Tube. Stanford, CA: Stanford University Press; 1997.

[13] Rheinberger, H-J. An Epistemology of the Concrete. Twentieth-Century Histories of Life. Durham & London: Duke University Press; 2010.

[14] Mol, A; Law, J. Complexities: an introduction. In: Law, J; Mol, A, editors. Complexities. Social Studies of Knowledge Practices. Durham & London: Duke University Press; 2002. p. 1–22.

[15] Bijker, WE. Of Bicycles, Bakelites, and Bulbs: Toward a Theory of Sociotechnical Change. Cambridge, MA: The MIT Press; 1997.

http://dx.doi.org/10.5772/66044

[16] Fuller, S. Science vs Religion. Intelligent Design and the Problem of Evolution. Cambridge, UK: Polity Press; 2007.

[17] Rudwick, JS. The Meaning of Fossils. Chicago: The University of Chicago Press; 1972.

[18] Gould, SJ. Ontogeny and Phylogeny. Cambridge, MA: The Belknap Press of Harvard University Press; 1977.

[19] Gould, SJ. Wonderful Life. The Burgess Shale and the Nature of History. New York and London: W.W. Norton & Company; 1989.

[20] Hopwood, N. Haeckel's Embryos. Images, Evolution and Fraud. Chicago: The University of Chicago Press; 2015.

[21] Gould, SJ. The Structure of Evolutionary Theory. Cambridge, MA and London: The Belknap Press of Harvard University Press; 2002.

[22] Jacobson, M. Development of specific neuronal connections. Science 1969; 163: p. 543–547.

[23] Hebb, DO. Essay on Mind. Hillsdale, NJ: Lawrence Erlbaum Associates, Publishers; 1980.

[24] Edelman, GM. Bright Air, Brilliant Fire. On the Matter of Mind. New York: Basic Books; 1992.

[25] Edelman, GM. Wider Than the Sky. The Phenomenal Gift of Consciousness. New Haven and London: Yale University Press; 2004.

[26] Edelman, GM. Second Nature. Brain Science and Human Knowledge. New Haven and London: Yale University Press; 2006.

[27] Varela, FJ; Thompson, E; Rosch, E. The Embodied Mind. Cognitive Science and Human Experience. Cambridge, MA: The MIT Press; 1993.

[28] Freeman, WJ. How Brains Make Up Their Minds. New York: Columbia University Press; 2000.

[29] Núñez, R; Freeman WJ, editors. Reclaiming Cognition: The Primacy of Action, Intention and Emotion. Thorverton, UK: Imprint Academic; 2000.

[30] Panksepp, J and Biven, L. The Archaeology of Mind: Neuroevolutionary Origins of Human Emotions. New York: W.W. Norton & Company; 2012.

[31] Damasio, A. Descartes' Error. Emotion, Reason, and the Human Brain. New York: Penguin Books; 1994.

[32] Damasio, A. The Feeling of What Happens. Body and Emotion in the Making of Consciousness. San Diego: Harcourt, Inc; 1999.

[33] Damasio, A. Looking for Spinoza. Joy, Sorrow and the Feeling Brain. San Diego: Harcourt, Inc; 2003.

[34] Damasio, A. Self Comes to Mind. Constructing the Conscious Brain. New York: Pantheon Books; 2010.

[35] Rose, N. Governing the will in a neurochemical age. In: Maasen, S and Sutter, B, editors. On Willing Selves. Neoliberal Politics vis-à-vis the Neuroscientific Challenge. Basingstoke and New York: Palgrave Macmillan Ltd.; 2007. p. 81–99.

[36] Woods, S. Respect for autonomy and palliative care. In: Have, H ten; Clark, D, editors. The Ethics of Palliative Care. European Perspectives. Buckingham and Philadelphia, PA: Open University Press; 1978. p. 145–165.

[37] Korein, J. editor. Brain Death: Interrelated Medical and Social Issues. New York: The New York Academy of Sciences; 1978.

[38] Wackers, G. Constructivist Medicine [thesis]. Ch. 5: Death in the skull. Re-negotiating the end of human life. Maastricht: Universitaire Pers Maastricht; 1994.

[39] Gervais, KG. Redefining Death. New Haven: Yale University Press; 1986.

[40] Veatch, RM. Whole brain, neocortical, and higher brain related concepts. In: Zaner, RM, editor. Death: Beyond Whole Brain Criteria. Dordrecht: Kluwer Academic Publishers; 1988. p. 171–186.

[41] Foster, C. Choosing Life, Choosing Death. The Tyranny of Autonomy in Medical Ethics and Law. Oxford: Hart Publishing; 2009.

[42] Rose, N and Abi-Rached, J. Neuro: The New Brain Sciences and the Management of the Mind. Princeton: Princeton University Press; 2013.

Chapter 11

Medical Ethics and Bedside Rationing in Low-Income Countries: Challenges and Opportunities

Lydia Kapiriri

Additional information is available at the end of the chapter

http://dx.doi.org/10.5772/65089

Abstract

There's evidence that implementing the four medical ethics principles may be challenging especially in low income country contexts with extreme resource scarcity and limited capacity to facilitate deliberations on the different ethical dilemmas. These challenges can partly be explained by the social, economic, and political contexts in which the decisions are made, as well as the limited time, training and guidance to facilitate ethical decision making. Based on current literature, and using the example of bedside rationing; this chapter synthesizes the challenges clinicians face when operationalizing the four principle; identifying the opportunities to address them. We suggest that clinicians' ability to implement the four principles are constrained by meso- and macro-level decision making as well as their lack of training, explicit guidelines, and peer support. To ameliorate this situation, current efforts to strengthen the clinicians' capacity to make ethical decisions should be complimented with developing of context relevant guidelines for ethical clinical decision making. The renewed global commitment to the sustainable development goals and universal healthcare coverage should be recognized as an opportunity to leverage resources and champion the integration of equity and justice as a core value in resource allocation at the bedside, meso-, macro- and global levels.

Keywords: principles of medical ethics, bedside rationing, low-income countries

1. Introduction

There is general acceptance of the four medical ethics principles of autonomy, beneficence, non-maleficence, and justice. These principles were developed in 1978 by the United States national commission for the protection of human subjects of health research involving human beings.

Originally, the commission identified three principles with corresponding (areas of application), namely respect for humans (informed consent); beneficence (assessment of risks and benefits); and justice (selection of human subjects) [1]. Since then, these principles form the basis of training in ethics for practitioners worldwide [2]. While these principles provide valuable guidance for practitioners, their applicability could be enhanced if they are contextualized, especially in the cases of low-income countries (LICs) where resources are extremely scarce [3]. Questions that may arise include the following: How do you obtain informed consent from a patient who is desperately in need of care and being unable to understand the treatment options—defers the decisions to the practitioner? How does a practitioner ensure that their "intervention" is of benefit to the patient when they do not have the resources to provide the best intervention? Furthermore, how do they ensure justice in the allocation of resources in the context of extreme resource constraints?

In contexts where resources are limited, rationing is recommended although it is contentious. There is a growing body of literature on bedside rationing (defined as the withholding of potentially beneficial treatment or medical procedure from a patient) [4–8]. While the literature highlights the general rationing challenges mainly related to the physicians' dual role as patients' advocates and society's gate keepers, this literature also clearly shows that in addition to the general challenges, health practitioners in low-income countries are faced with many unique ethical challenges in their practice, with dire consequences for both themselves and their patients [7–9]. This could partly be explained by (i) the context in which they practice which is characterized with low literacy rates, poor infrastructure, poverty, marked social, health inequalities; limited government commitment to equitable health systems and the global health context; (ii) their training in medical ethics and the availability of explicit guidelines for decision making; and (iii) external pressures.

This chapter highlights the challenges practitioners working in low-income countries (LICs) face when operationalizing the general principles of medical ethics. We expound on the principle of justice, using the example of bedside rationing, and discuss how these decisions are made and their ethical implications and challenges. In next sections, we will first provide an overview of the LIC contexts, identifying the factors that could influence ethical decision making; second, we will discuss the challenges related to operationalizing the four principles; and third, we will use the case of bedside rationing to expound on the justice principle and the related challenges. In the last section, we will discuss the current opportunities and potential interventions that would alleviate the challenges faced by practitioners in low-income countries.

2. The context of medical ethics and bedside rationing in low-income countries

According to the World Health Organization, low-income countries share similar characteristics of high levels of population poverty, poor physical infrastructure, lower literacy levels, and weak health systems as illustrated in the **Table 1**.

http://dx.doi.org/10.5772/65089

Indicator	Value
Gross domestic product	$397.8
Population leaving on less than $1.90/day	47.2%
Literacy level	57%
Rural population	70%
Health expenditure per capita	$37
Hospital beds/1000 population	0.8
Physician/1000 population	0.1
Births attended by trained staff	50%
Life expectancy at birth	61 years
Infant mortality rate/1000 live births	53

Source: http://data.worldbank.org/income-level/LIC (Accessed June, 2016).

Table 1. Socioeconomic characteristics of low-income countries.

As demonstrated in **Table 1**, the limited resources available to the health system in most low-income countries ($37 per capita compared to the >$1000 for most of the OECD countries) have resulted in ad hoc availability of medical supplies and limited and often poorly motivated health workers. The dismal human resources/population ratios means that poorly motivated health workers have a very high patient load [10], which impacts the amount of time a clinician can spend with a patient and their quality of care. This situation is worse for the poor, rural populations where due to a lack of physical infrastructure and poor access to opportunities, health workers are often reluctant to live and work in these areas [10].

Most LIC governments, recognizing these problems, have embarked on re-structuring their health systems in order to improve population access to quality healthcare services. This has been achieved through the commitment to universal health coverage and the re-orientation of services to primary health care and building and strengthening the primary care network. There have also been attempts to strengthen the referral system to ensure that only referred patients are sent to hospitals [11]. However, in many cases, the referral system has not functioned as planned, due to the limited and often ad hoc availability of medical supplies at the lower levels. Patients end up crowding secondary and tertiary level units where they are likely to be seen by a specialist and are more likely to receive treatment [12].

The resource constraints experienced by health systems in LICs are a reflection of these countries' low GDP and their limited per capita health expenditure. Some scholars have strongly argued that since governments are unable to meet their commitment to publicly finance the health systems, alternative financing mechanisms should be sought [13]. However, while health insurance has worked in places where most of the population is formally employed, it has not worked well in most LICs where 70% of the population is rural and employed in the informal sector. Organizing and managing such a system has proved to be more costly than its benefits. There have also been attempts to introduce user fees; however, in several contexts, this has also failed due to the financial barrier it introduces, for especially the poor. Hence, in the interest of equitable access, this has been abandoned [14].

Within this context, medical education emphasizes and urges providers to uphold the standard four principles of medical ethics. However, the context of extreme poverty, low literacy levels, social-economic and health inequities, poor infrastructure, and weak health systems may present challenges for clinicians who may wish to operationalize these principles. We explore these in detail in the next section.

3. The principles of medical ethics and their operationalization in low-income countries

Medical ethics hinge on four principles, namely autonomy, non-maleficence, beneficence, and justice. We discuss the challenges related to operationalization these principles in LICs.

3.1. Autonomy

According to this principle, patients should have the freedom in thought, intention, and action to make informed choices with regard to their treatment. They should be provided with adequate information on the risks and benefits of an intervention and should not be influenced or coerced into making a particular choice. Individuals should also be able to understand the consequences of his/her actions [15]. Inherent to this principle is the assumption that the patient has the capacity, understands, and is capable of acting independently without outside influence and that they want to engage in the process of determining their treatment. A secondary assumption is that the patient wants to independently make these decisions.

While these assumptions might hold in some contexts, several contextual factors limit the ability of LIC clinicians to uphold this principle. The high clinician:patient ratio and low literacy levels may make it difficult for clinicians to adequately explain treatment options to patients and may make it difficult for patients to adequately understand the information that is provided in order for them to make independent, informed choices. Moreover, in some cultures in Africa, it is unacceptable for sick people to be "burdened" with making decisions about their health, neither do they believe in individualized decision making [2]. In other instances where gender inequities are prevalent, women are often prohibited from making decisions independent of their husbands. As such, patients' ability to choose family and cultural interests and the physician's obligations in relationship to autonomy can, at times, be in conflict, leaving physicians with the challenge of how to handle these different interests while still upholding the ethical principles [16].

While the discussion of autonomy often exclusively focuses on the patient, here we, of necessity also reflect on clinicians' autonomy in medical decision making. While it is often assumed that clinicians have the autonomy to make independent decisions in their patients' best interests, this assumption is flawed within the LIC context. First, most LICs provide an essential drug list which might limit clinicians' autonomy in deciding the drugs they can prescribe for their patients [17]. Second, a lack of infrastructure and other medical technologies constrain clinicians' autonomy to decide when and how to treat their patients. Finally, due to low salaries, clinicians may be liable to coercion by pharmaceutical companies to use the drugs and technologies they want to promote. On the other hand, due to the lack of regular supply of drugs and poverty among patients, some clinicians, in the quest to do what is best for their patients in the face of scarcity, may be forced to use the samples of drugs introduced by the pharmaceutical companies, since the samples are free. While this may seem reasonable at the time, in that the patient gets some treatment at the time; it raises questions with regard to whether or not the patient is able to obtain the full course of the drug (which can be costly especially if newly introduced), and if they are unable to, then the benefits are short lived [18, 19].

3.2. Non-maleficence

According to this principle, a clinician should not intentionally cause harm to a patient or society either by commission or by omission. They should also do whatever is possible to prevent any harm to the patient [20]. Clinicians should have the competency to provide the appropriate standard of care and the authority and freedom to make clinical decisions [1].

Medical schools in LICs endeavor to provide the best possible education to their trainees, equip them with the knowledge and competencies they need to practice medicine, and provide the appropriate standard of care [21]. However, an extreme lack of resources, especially drugs and medical supplies, often forces clinicians to make decisions (with regard to the choice of treatment or medical procedure), which may not adhere to the standard of care and could potentially be harmful to patients and society. For example, in some cases, clinicians are unable to ensure that the patients get the full course of the necessary antibiotics, which puts patients at risk of developing drug resistance, which is detrimental to both patients and society [22]. The dilemma faced by these clinicians then is whether to defend a patient's or society's interest. Withholding limited drugs from patients who are unlikely to be able to afford to complete the prescribed treatment would prevent these patients from developing drug resistance, which is good for both the patient and society. However, the patient may not perceive it as beneficial in that moment. The other option would be to honor the patient's request and provide whatever drugs that are available, although inadequate, to the patient, even if the practitioner knows that the patient would not afford to pay for the rest of the treatment. While this would appease the patient, it may be detrimental to both the patient and society in the long run, when such patients develop drug resistance. Unfortunately, some clinicians, due to the various reasons discussed above, are more likely to take the second option [7].

3.3. Beneficence

This principle, sometimes perceived to be the opposite of maleficence, is based on minimizing risks and maximizing benefits. At the bedside, it relates to the clinician's commitment to doing what is right, an act of charity, mercy, or kindness to another person. In this context, it is the moral obligation for clinicians to ensure that any procedures or treatment given to the patient is intended to benefit the patient and minimize risk [23]. This requires clinicians to have the appropriate skills and knowledge, the ability to understand and assess the individual patient's circumstances and to assess the risks and benefits of the available treatment options, in order for them to confidently determine that the treatment option would definitely benefit the patient and cause them no harm [23, 24].

Similar to the previous principles, beneficence assumes that the clinicians are in complete control of making and executing any clinical decisions that they deem beneficial to their patients. However, while they may have the expertise, intention, and commitment to act morally, out of charity, mercy, and kindness towards their patients to ensure that their patients benefit, a general lack of resources, and the inability of the patients to access the treatment of choice means that clinicians' commitment to benefit the patient is not always possible [7, 8].

Another challenge relates to the ability of the clinicians to confidently determine that the available treatment would benefit (and not harm) the patients. This necessitates ensuring that the benefits of any procedure or treatment outweigh the risks. However, most of LIC clinicians are not always able to assess this; it may not even be possible to accomplish this at the national level. Such situations result from a lack of the capacity at the national or hospital laboratories to test procedures as they are developed in order to assess their applicability within the local context.

Some of these challenges could be explained by reflecting on the global research and development for new drugs. Due to low research capacity in most LICs, many of the drugs are developed and tested in high- or middle-income countries [25]. The context of the drug development differs from the LIC context where most patients are likely to have comorbidities and may respond to the treatment differently [26]. While having well-functioning laboratories and research capacity within LICs would help clinicians to assess the effectiveness of these drugs within this context; in many cases, national laboratories are ill-equipped and lack the capacity to conduct these complex assessments. Moreover, the rate at which new drugs and technologies are developed does not give clinicians adequate time to review and evaluate information related to one new drug before another drug or technology is introduced and promoted [27]. For these reasons, clinicians are forced to make decisions with regard to adopting new drugs without complete knowledge of how the patients may react to the treatment and are therefore unable to say definitely that they are fulfilling the beneficence principle.

3.4. Justice

This principle relates to fairness in the allocation of health resources including drugs, practitioner's time, and health facility procedures. Sometimes this is interpreted in terms of distributing resources: either equally in society/or to individuals; or according to need, share, effort, contribution, merit, and free-market exchanges [28]. While the distribution of health resources occurs at the different levels of decision making, this paper focuses primarily on the distribution of resources at the bedside. We expand on the challenges related to this principle in relationship to clinical decision making at the bedside in LICs. The discussion in this section also draws on the prior discussion of the three principles of medical ethics.

http://dx.doi.org/10.5772/65089

4. The challenges related to the principle of justice: a look at allocating resources at the bedside in LICs

In order for us to understand the challenges related to bedside rationing in low-income countries, we will start by reviewing some of the literature on the rationing of treatment in low-income countries, then present cases of the rationing process of drugs from the literature, discussing the challenges. Finally, we propose some ways to address these challenges.

There is meager literature on bedside rationing in LICs. Kapiriri and Martin, 2007 [8] describe bedside rationing in a hospital in Uganda. This study provides a detailed decision making process that clinicians follow when making medical decisions—which were often related to setting limits to treatment access. Decisions about who receives first-line treatment, which patient is seen first and who gets to go the operating room were discussed. This study reported that clinicians considered both acceptable and non-acceptable criteria when making these decisions. The decisions were constrained by decisions made at the national and global levels. In another study, Defaye et al. [9] conducted a national survey of clinicians, the majority whom reported to frequently make rationing decisions in relationship to referrals for surgery, ICU, and prescription drugs. Many had witnessed the adverse effects of bedside rationing including death and disabilities. Both these cases documented physicians experiencing moral distress as a result of making the rationing decisions.

This literature demonstrated that while the clinicians had the capacity to make definite diagnoses and prescribe the first-line treatment for the patients, the availability of the prescribed treatment or procedure forces clinicians to consider if the patients are able to purchase the first-line treatment, if not, they receive the second-line treatment. While these decisions are sometimes made with the patients, sometimes, they are independently made and not guided by an explicit criteria or guidelines [7, 8, 18]. These findings are similar to those of a study conducted in India on rationing of care for neonates [29].

The challenges that clinicians face seem to be consistent across countries. We summarize the above challenges in **Table 2**.

Some challenges were common to all the three cases: the context of extreme resource scarcity, lack/inadequacy of explicit guidelines, the challenge of having to forego the first-line treatment due to scarcity of resources, and the challenge to choose what patients to withhold treatment from. There were practitioner-related challenges which were also common to the three contexts: the use of unacceptable criteria, the dual role of practitioners, witnessing the consequences of the rationing decisions, and the resulting moral distress. Lack of education resulting in inability to discuss the decisions with patients and the patients' differing the decisions to the clinicians were common to the three cases.

Only three challenges were identified in only one context. Disagreements among health professionals with regard to how to ration care in the Ethiopia study, lack of forum for discussing ethical issues in the India case study, and the lack of credible in the Uganda case study. The rest of the challenges were common to at least two of the case studies.

While structural factors: poverty, lack of education, lack of medical supplies present insurmountable challenges, some of the challenges such as lack of guidelines for decision making, lack of proper training in making fair rationing decision are modifiable and present opportunities for improvement [7, 8, 18, 29].

Types of challenges	Ethiopia	India	Uganda
Contextual challenges			
Resource scarcity	X	X	X
Patients' socioeconomic status: poverty, education	X	X	X
Decision making process-related challenges			
Lack of training in resource allocation ethics	X		X
Lack/inadequacy of guidelines	X	X	X
Disagreements among professionals	X		
Foregoing first choice treatment	X	X	X
Choice of patients to restrict access to treatment	X	X	X
Balancing patient and family interests	X	X	
Lack of credible evidence			X
No forum for discussion ethical issues		X	
Practitioner-related challenges			
Perceived lack of control of the decision yet held responsible		X	X
Feelings of helplessness and incompetence		X	X
Use of unacceptable criteria and how to deal with it	X	X	X
Dual role of the physicians	X	X	X
Witnessing and feeling responsible for the consequences of rationing for the patient	X	X	X
Moral distress	X	X	X
Patient-related challenges			
Inability to communicate due to lack of knowledge	X	X	X
Differing decisions to doctors		X	X
Excessive demand by "able" patients		X	X

Table 2. Sample of documented bedside rationing challenges in low-income countries.

5. Discussion

Several interventions are proposed to deal with the above challenges and facilitate the implementation of the four principles of medical ethics in LICs. We discuss these in detail.

5.1. Opportunities and interventions at the bedside

i. *Contextualize the principles of medical ethics*: While bioethics principles are accepted globally and used as basis for medical training, the realities in LICs make it impossible for clinicians to implement these principles as postulated. As discussed above, the social, economic, and cultural contexts where the medical decisions are made constrain the application of these principles. Given the critical relevance of ensuring that clinical practice is ethical even in LICs demands that medical ethics remain at the core of medical training. However, the above challenges would require that the principles are contextualized [2]. This would entail not diluting them, but ensuing that they are relevant and can provide meaningful guidance to the clinicians who should apply them. This could be achieved by introducing the ideal principle but supplementing these with relevant case scenarios from the local context, in order to ensure that trainees engage and understand how these principles can be applied within their context and the related dilemmas. This would enable students to think through the different challenges they might face and develop context-sensitive strategies.

ii. *Develop explicit medical ethics and rationing guidelines*: The literature to date highlights that practitioners in LICs lack explicit guidelines for use when making limit setting decisions. This does not seem to be an integral part of their training in bioethics. There is hence, need for LIC medical schools, in conjunction with ethicists to develop explicit guidelines for clinicians. This would not only ensure consistence in their decision making but would also contribute to reducing the clinicians' moral distress if they know that their decisions would be supported by other clinicians. Such guidelines should include clear decision making flow diagrams, with explicit criteria (both medical and otherwise) which should be developed with input from the public [7, 8, 18]. The guidelines should also include the principles of procedural justice in resource allocation. This entails ensuring that the criteria used is relevant to the decisions, that the decisions and criteria are publicized, that there are provisions for appealing and revising the decision and that there are voluntary mechanisms to ensure adherence to this process [28]. These guidelines would be an integral addition to the existing medical ethics courses in the schools of medicine.

iii. *Strengthen in service peer support systems*: In addition to training and developing decision making guidelines, there is need for regular clinical discussions on the ethical challenges faced by practitioners [29]. This could be part of the routine clinical departmental meetings that are regularly convened to review patient care. Facilitating dialogue about the ethical challenges and how to best handle them would serve to mitigate the moral distress that clinicians experience. Furthermore, such a forum could also be used to re-emphasize the consistent use of the guidelines and their revision where necessary.

5.2. Opportunities and interventions at the health institution level

i. *Support the efforts at the micro-level*: Since the clinicians work under the leadership of health institutions, it is critical for micro-level interventions to be supported by the institution leadership. For example, the guidelines developed to guide clinicians in ethical decision making with regard to resource allocation could be formalized for use across the clinical departments.

ii. *Lead by example*: The institutional administrators should lead by example by implementing fair priority setting processes at their level of decision making [28]. In order to facilitate this, the institutions should develop explicit priority setting processes, with criteria which is developed in conjunction with clinicians. These criteria should be published with the institution and debated, it should also reflect the relevant criteria identified at the micro-level. Once resource allocation decisions are made, they could be disseminated through an institutional newsletter, where available and also posted on the institutional webpages. The management of the institution should also establish clear mechanisms for appealing the decisions. While many institutions have used suggestion boxes, these have been found to be ineffective [18]. A designated office that handles complaints related to the allocation of resources may be more accessible to the public. Information about this office should be publicized. This internal office could also be responsible for ensuring that the conditions for fair priority setting are met within the institution.

iii. *Capacity building*: There is consistent observation that decisions made at the meso-level and macro-level constrain the bedside rationing decisions [18, 29]. If the hospitals do not facilitate shared decision making and transparency in their decision making, it is unlikely that the clinicians will abide with the proposed limit setting decisions. Furthermore, availability of resources at this level determines the resources that clinicians have for use. In many cases, hospital managers are not clinicians and may not necessarily be trained in medical ethics or in medical resource allocation principles [7, 8, 29]. In cases where officers have not received this critical training, there should be in-service training of all institutional leaders in the principles of justice in the allocation of resources. An example of such training has been introduced in Ethiopia where clinicians and leaders in health are trained in medical ethics, including justice in resource allocation [30]. This could be a model that could be duplicated in other LICs in order to strengthen the capacity of hospital managers and clinicians to make ethical resource allocation decisions.

5.3. Opportunities and interventions at the national level

i. *Political will and commitment*: While it is important that clinicians and hospital managers receive training and facilitate fair resource allocation decision making, these would not address the LIC contextual factors that constrain their decision making. Concurrent efforts at the national level need to focus on:

a. Increasing the health sector budget: While it is understood that LICs have limited resources, countries need to commit to at least meet the Abuja declaration on government funding of the health sector [31]. While LICs have, in the past, depended on donors to support their health sectors, this ought not to be the norm since it is non-sustainable and has the repercussions of having national priorities

http://dx.doi.org/10.5772/65089

influenced by non-state stakeholders. Priority setting, if done well, would help governments focus on what is feasible and sustainable within their budgets.

b. Strengthening health system and infrastructure and allocating adequate resources to at the very minimum deliver the essential health services. While most LICs have defined a basic healthcare package, not many have been able to deliver all the basic services. Sited limitations include a mismatch between the package and both financial and human resources available to deliver it [32]. The limited resources often mean that clinicians do not have the basic drugs and diagnostics to treat the patients effectively. It is hence unlikely that the challenges faced by clinicians can be successfully mitigated without addressing these contextual factors. The recent committed to achieving Universal Health Care coverage by most of the low-income countries provides a unique opportunity for strengthening health systems; however, this needs to include concurrent efforts to ensure that the package serves those who are most in need of the services; the use ethical principles to allocate the available resources, with specific attention to equity in access is critical [33].

c. Improving decision making at the national level: National level priority setting in many LIC health system has been described as ad hoc due to lack of explicit processes and the resources to implement the identified priorities. This results in a high potential for national priorities to be overlooked during implementation [34]. While there are national level processes where priorities are identified, they are often aligned with the essential service package (discussed above), which governments are unable to fund. Governments should use the universal health access commitment to identify realistic services (within the available resources) services that their populations, especially the most vulnerable should have access. Similar to the meso-level, the existing decision making processes need to be strengthened by the following: establishing explicit fair priority setting process which is participatory, transparent and accountable. The public should inform and be informed about this process with the publicity of both the decisions and criteria [27, 33]. Governments also need to devote resources to addressing the key non-health system determinants of health that impact the population access to health and health care such as education and poverty [35].

ii. *Facilitate public awareness and education*: The public is often unaware of the need to ration care. This is worsened by the politicians' reluctance to discuss the realities of lack of medical supplies and human resources in the health institutions with the general population [36]. In order to empower patients (and the public), health ministry's should facilitate open discussions about the limited resources and the need to ration, at the different levels. Public values and preferences with regard to priority setting should be elicited and be used as input to priority setting processes. While this might be constrained by the low population literacy levels, presenters of such information should use innovative strategies, for example, using the balanced sheet, charts, and diagrams, to ensure that the information is accessible to the public [37,

38]. The existing decentralized political and governing structures provide an opportunity for public engagement since one of the purposes of decentralization is to take decision making closer to the people and facilitating public engagement in decision making.

5.4. Opportunities and interventions at the global level

The renewed global commitment to the Sustainable Development Goals, with health and well-being as goal 3, presents an opportunity for leveraging resources for LIC health systems since most of the health systems within low-income countries are supported development assistance partners [39]. This interest and commitment by development partners to invest in health interest should be meaningfully exploited. Some of these resources could be committed to strengthening and supporting the priority setting and ethics infrastructure within LICs.

Global level stakeholders also need to observe the key guiding principles of fair decision making when deciding where to invest their resources. These processes should consider information and priorities within the countries they may wish to support. Furthermore, global partnerships should aim at realizing equitable access to drugs by ensuring that their research and development investments are made fairly reflecting equity. Once the drugs and diagnostics are developed, there should be frameworks to ensure that LICs can access the new innovations. The challenge for the WHO is to strengthen such existing frameworks [40]. However, the ethical principles especially conflict of interest should be carefully monitored in such partnerships.

6. Conclusions

The context of decision making in low-income countries constrain clinicians' ability to implement the principles of medical ethics and especially justice when caring for their patients. Efforts that have focused entirely on strengthening the capacity of clinicians in medical ethics, without addressing the contextual factors that constrain these decisions are limited. There is need for concurrent efforts to focus at the meso-, macro- and global levels so as to ensure that the context where practitioners make medical decisions is conducive for ethical decision making.

Author details

Lydia Kapiriri

Address all correspondence to: kapirir@mcmaster.ca

Department of Health, Aging and Society, McMaster University, Hamilton, ON, Canada

References

[1] The United States Department of Health, Education and Welfare; The Belmont Report. https://videocast.nih.gov/pdf/ohrp_appendix_belmont_report_vol_2.pdf Accessed July 2016.

[2] Buchanan A, Brock DW. The concept of informed consent In: Fisher ed. Biomedical Ethics; A Canadian Focus, 2nd edition. Oxford University Press. 2013;26–29.

[3] Olweny C. Bioethics in developing countries: ethics of scarcity and sacrifice. J Med Ethics. 1994;20(3):169–174.

[4] Hurst SA, Slowther AM, Forde R, Pegoraro R, Reiter-Theil S, Perrier A, Garrett-Mayer E, Danis M. Prevalence and determinants of physician bedside rationing. J Gen Intern Med. 2006;21(11):1138–1143.

[5] Bloche MG, Jungman ER. The "R" word. J Contemp Health Law Policy. 2002;18:633–639.

[6] Askin WJ. Bedside rationing. CMAJ. 2002;166:711.

[7] Weinstein MC. Should physicians be gatekeepers of medical resources? J Med Ethics. 2001;27:268–274.

[8] Kapiriri L, Martin DK. Bedside rationing by health practitioners: a case study in a Ugandan hospital. Med Decis Making. 2007;27(1):44–52.

[9] Defaye FB, Desalegn D, Danis M, Hurst S, Berhane, Norheim OF, Miljeteig I. A survey of Ethiopian physicians' experiences of bedside rationing: extensive resource scarcity, tough decisions and adverse consequences. BMC Health Serv Res. 2015, 15:467.

[10] Coulter A, Ham C. The Global Challenge of Health Care Rationing. Buckingham: Open University Press. 2000.

[11] World Health Organization 2010. The World health report 2010. Health systems financing: the path to universal coverage. http://www.who.int/whr/2010/en/

[12] Rodin J, de Ferranti D. Universal health coverage: the third global health transition? The Lancet 2012;380(9845):861–862.

[13] Nakahara S, Saint S, Sann S, Ichikawa M, Kimura A, Eng L, Yoshida K. Exploring referral systems for injured patients in low-income countries: a case study from Cambodia. Health Policy Plan. 2010;25(4):319–327.

[14] Kutzin J, Yip W, Cashin C. Alternative financing strategies for universal health coverage. In: World Scientific Handbook of Global Health Economics and Public Policy. 31 2016;267-309 R. M. Scheffler (UC Berkeley) Ed. http://www.worldscientific.com/doi/pdf/10.1142/9789813140493_0005

[15] WHO, 2015. Workload indicators of staffing need (WISN): selected country implementation experiences. Accessed at: http://apps.who.int/iris/bitstream/10665/205943/1/9789241510059_eng.pdf?Ua=1

[16] Faden RR, Beauchamp TL. The concept of informed consent. In: Fisher ed. Biomedical Ethics; A Canadian Focus, 2nd edition. Oxford University Press. 2013;44–46 (15:467).

[17] Mappes TA, Zambaty JS. Patient choices, family interests, and physician obligations. Kennedy Inst Ethics J. 1994;4(1):27–46.

[18] Sharma S, Kh R, Chaudhury RR. Attitude and opinion towards essential medicine formulary. Indian J Pharmacol. 2010;42(3):150–152.

[19] Kapiriri L, Martin DK. Priority setting in developing countries health care institutions: the case of a Ugandan hospital. BMC Health Serv Res. 2006;6:127.

[20] Kapiriri L. Ethical challenges in clinical decision-making in the era of new technologies: experiences from low income countries. Health Policy Technol. 2015;4(3):256–262.

[21] Omonzejele PF. Obligation of non-maleficence: moral dilemma in physician-patient relationship. JMBR 2005;4(1):22–30.

[22] Cueto J Jr, Burch VC, Adnan NA, Afolabi BB, Ismail Z, Jafri W, Olapade-Olaopa EO, Otieno-Nyunya B, Supe A, Togoo A, Vargas AL, Wasserman E, Morahan PS, Burdick W, Gary N. Accreditation of undergraduate medical training programs: practices in nine developing countries as compared with the United States. Educ Health (Abingdon). 2006;19(2):207–222.

[23] Reardon S. Antibiotic resistance sweeping developing world; Bacteria are increasingly dodging extermination as drug availability outpaces regulation. Nature. 2014;509:141–142.

[24] Kinsinger FS. Beneficence and the professional's moral imperative. J Chiropr Humanit. 2009;16(1):44–46.

[25] Kapiriri L, Lavery J, Singer PA, Mshinda H, Babiuk L, Daar AS. The case for conducting first-in-human (phase 0 and phase 1) clinical trials in low and middle income countries. BMC Public Health. 2011;11:811.

[26] Glickman SW, McHutchison JG, Peterson DE, Cairns CB, Harrington RA, Califf RM, Schulman K. Ethical and scientific implications of the globalization of clinical research. NEJM Sounding Board. 2009;360(8):816–823.

[27] Emanuel EJ. The paradox of exploitation: the poor exploiting the rich. In: Lavery JV, Grady C, Wahl ER, Emanuel EJ (eds) Ethical Issues in International Biomedical Research: A Casebook. New York: Oxford University Press. 2007:189–195.

[28] Entwistle VA, Sheldon TA, Swoden A, Watt IS. Evidence-informed patient choice: Practical issues of involving patients in decisions about health care technologies. Int J Technol Assess Health Care. 1998;14(2):212–225.

[29] Daniels N, Sabin JE. Setting Limits Fairly: Can We Learn to Share Medical Resources? Oxford, UK: Oxford University Press. 2002;25–46.

[30] Miljeteig I, Norheim OF. My job is to keep him alive, but what about his brother and sister? How Indian doctors experience ethical dilemmas in neonatal medicine. Dev World Bioeth. 2006;6(1):23–32.

[31] Training of trainers Ethiopia: http://www.uib.no/en/rg/globpri/83274/training-trainers-medical-ethics-tot-me-ethiopia. Accessed July, 2016.

[32] WHO, 2011; Abuja Declaration http://www.who.int/healthsystems/publications/abuja_report_aug_2011.pdf?ua=1

[33] Ssengooba F. Uganda's minimum health care package: rationing within the minimum? *Health Policy and Development*. 2004, 2(1).

[34] Ottersen T, Norheim OF, Bona M Chitah BM, Cookson R, Daniels N, Defaye FB et al. (2014) Making fair choices on the path to universal health coverage. Final Report of the WHO Consultative Group on Equity and Universal Health Coverage. Geneva: World Health Organization.

[35] Kapiriri L. Priority setting in low income countries: the roles and legitimacy of development assistance partners. Public Health Ethics. 2012;1(5):67–80.

[36] WHO, 2005 http://www.who.int/social_determinants/thecommission/finalreport/closethegap_how/en/index3.htmlSocial determinants

[37] Kapiriri L, Randal G, Martin DK. Health care rationing and professional autonomy: the case of cardiac care in Ontario. Open Health Serv Policy J. 2009;2(8):34–41.

[38] Makundi E, Kapiriri L, Norheim OF. Combining evidence and values in priority setting: testing the balance sheet method in a low-income country. BMC Health Serv Res. 2008;7(1):152.

[39] WHO, 2016a. Meeting the need for treatment: the initiatives. http://www.who.int/bulletin/volumes/84/5/news30506/en/

[40] WHO, 2016b. http://www.who.int/mediacentre/events/meetings/2015/un-sustainable-development-summit/en/

Chapter 12

'Assisted Dying': A View of the Legal, Social, Ethical and Clinical Perspectives

Ilora Gillian Finlay of Llandaff

Additional information is available at the end of the chapter

http://dx.doi.org/10.5772/65908

Abstract

Discussion of legislation of physician-assisted suicide and euthanasia, often euphemistically called 'assisted dying', frequently focuses on individual cases promoted by campaigners as the reason that the law to licence doctors to supply lethal drugs to patients requesting them should change under certain conditions. But such legislation has wider consequences that simply for a handful of cases, as the relentlessly increasing numbers of such deaths have shown.

Keywords: assisted dying, physician assisted suicide, euthanasia, religion, legalisation, Parliament

1. Introduction

Current law in the United Kingdom and in most parts of the world is crystal clear. There is an absolute prohibition on killing another person, and it is illegal to help someone take their own life. But society's prohibition on assisting suicide has become eroded. Fear of dying has been publicised, with premature death being portrayed as a preferable option to a natural death with all care given. And there have been challenges to the clinical boundaries of care through court challenges about the cessation of interventions in some patients in minimally conscious states.

Against this backdrop campaigners for assisted suicide and euthanasia have mounted large media campaigns, which have gained some traction in increasingly secular and utilitarian societies. A general perception of 'better off dead' has fallen into common parlance, and yet the reality of death has become increasingly unfamiliar to people as the majority of deaths occur in hospitals out of sight of family and friends. Additionally, there has been relatively little publicity about improvements end-of-life care.

Assisted dying is a recently coined term that covers physician-assisted suicide (PAS) and physician administered euthanasia (PAE). Wherever such acts have been legalised, they all aim to achieve the same goal: they licence doctors to prescribe lethal drugs to people who request them, with certain criteria being stipulated. But it becomes important to unpick exactly what is meant by the broad blanket term 'assisted dying'. Various laws or proposed laws around the world have used the phrase in their title.

In 1998, the US State of Oregon enacted the 'Death with Dignity Act' (DWDA) [1]. California's recent 'End of Life Option Act' [2], and now Canada's 'Medical Assistance in Dying Act' imply that without the option of assisted suicide or euthanasia, death will be far worse than if life is abruptly foreshortened [3]. The Dutch in 2001, with commendable honesty in its title, passed their 'Termination of life on Request and Assisted Suicide Act' [4]–the title of the legislation describes exactly what the legislation does.

The problem with titles that are themselves euphemisms is that the public are misled, as are politicians, over what exactly legislation can and cannot do and over potential unintended consequences of it.

One difficulty is that criteria in legislation, intended to define clearly who is or is not eligible for such lethal drugs, are of themselves open to wide interpretation and some are based on flawed assumptions.

However, before the law is changed, it is necessary to ask whether there is compelling evidence that the law needs to be changed and that the benefits of such a change outweigh the adverse consequences of legal change. Such legislation alters the focus and ethics of clinical decision-making, it alters society's approach to those who for whatever reason are seeking suicide and it alters a fundamental moral code in society that one person should not deliberately bring about the death of another.

2. The law

In most countries in the world, there is a clear prohibition on assisting suicide.

Suicide is regarded as a very grave matter. Despair that drives someone to suicide requires a response of trying to support that person, as embodied in suicide prevention policies. These laws recognise the duty on all in society and also recognise the vulnerability of the profoundly depressed person with suicidal ideation to coercive pressures on them, however subtle, to think they would indeed be 'better off dead'. Linked to that is the compassionate approach to the person who attempts suicide and fails–suicide per se is not a criminal offence.

Historically, this was not the case and suicide itself was viewed as self-murder, carrying with it the opprobrium of society. However, by 1961, the UK, as well as many other countries, had recognised the inhumanity of such a law that left those who survived attempting suicide to be potentially prosecuted for the actions their despair had driven them to. And it also recognised that society was failing such people if it did not tackle the source of despair, usually severe psychiatric depression, and that suicide prevention policies needed to be strengthened.

http://dx.doi.org/10.5772/65908

This led to the 1961 Suicide Act [5], which decriminalised suicide but deliberately did not legalise it, because legalisation would have brought the stamp of approval on suicide itself. Parliament was very clear that suicide should remain a very grave action and one that society has a duty to do all possible to prevent such actions, and respect life itself. But Parliament also recognised, in decriminalizing suicide, that criminal sanctions were no way to treat and support those in urgent need of psychiatric help.

By the end of the twentieth century, new sources of coercion to suicide had emerged as internet suicide sites goaded vulnerable young people to commit suicide; some such sites even had voyeuristic onlookers as the person died [6]. To meet these changing societal pressures, the Coroners and Justice Act in 2009 [7] amended the wording of the 1961 Suicide Act, to broaden the scope of the offence from that of 'aiding and abetting suicide' to the wider offence of 'encouraging or assisting suicide'.

Although this offence carries with it a maximum sentence of 14 years, it carries no minimum sentence that the courts must apply if a person is found guilty. Prosecutions for assisting suicide also require the agreement of the Director of Public Prosecutions (DPP) to proceed. The process of deciding whether or not to prosecute a person for assisting a suicide was clarified in 2010, when the then Director of Public Prosecutions was required by the House of Lords acting in its previous judicial role, which is now the Supreme Court to publish his guidance on prosecuting such acts. This clarified that there is a two-stage test that must be satisfied for a prosecution to proceed: first, there must evidence that the suspect did an act capable of and intended to encourage or assist suicide and second that such a prosecution is in the public interest. The factors which may tend towards prosecution and those tending to mitigate against prosecution were published at that time and were further clarified in 2014 [8]. Since then, the number of cases referred to the DPP has remained small with 124 cases referred to the DPP in the 7 years from 2009 to 2016 and the law continues to act as an effective deterrent to such actions. Because the two-stage test must be fulfilled, among these cases 102 did not proceed, because 24 were withdrawn by the police and in 78 cases the Crown Prosecution Service decided not to proceed. In some of these the assistance was given out of extreme compassion after trying all avenues to dissuade the person from their suicidal course of action. Among the remaining cases, six cases proceeded to prosecution for homicide or similar serious crime [9].

This prosecutorial discretion is not unique to the law on assisting suicide; it is the way the law works. Non-prosecuting does not mean the law is not working, it means the law sets clear criteria, people know what they are and when the law is breached the circumstances pertaining to the situation will be examined–in other words, the law has a stern face but an understanding heart. Take another example, a person stealing at gunpoint for personal gain can expect to be dealt with harshly, whereas a mother stealing food to feed her hungry child might well be dealt with quite differently, but no one is proposing a law to exempt certain types of theft from potential prosecution.

The law itself lays down a clear black and white line about what is acceptable and what is not; it sends social messages. It is the interpretation of the law that is then circumstance specific.

This messaging is evident in the numbers of deaths from 'assisted dying' seen in jurisdictions that have legalised such practices and where figures suggest a normalisation of the practice is

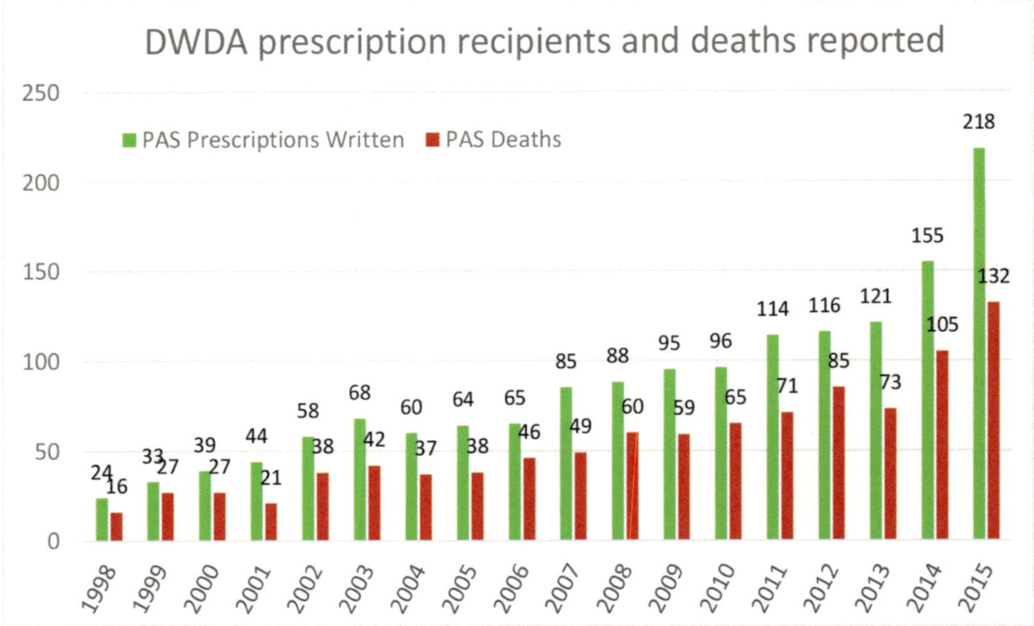

Figure 1. Oregon's number of lethal prescriptions issued and deaths from lethal drugs recorded as being under the Death with Dignity Act [10].

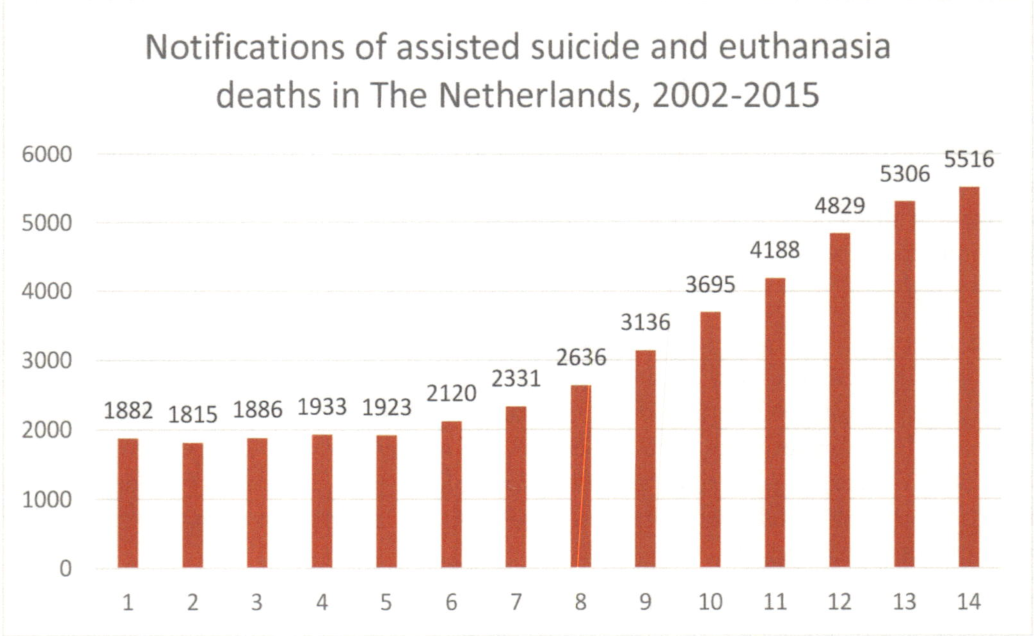

Figure 2. Deaths reported as deaths under the Termination of Life on Request and Assisted Suicide legislation in the Netherlands [11].

http://dx.doi.org/10.5772/65908

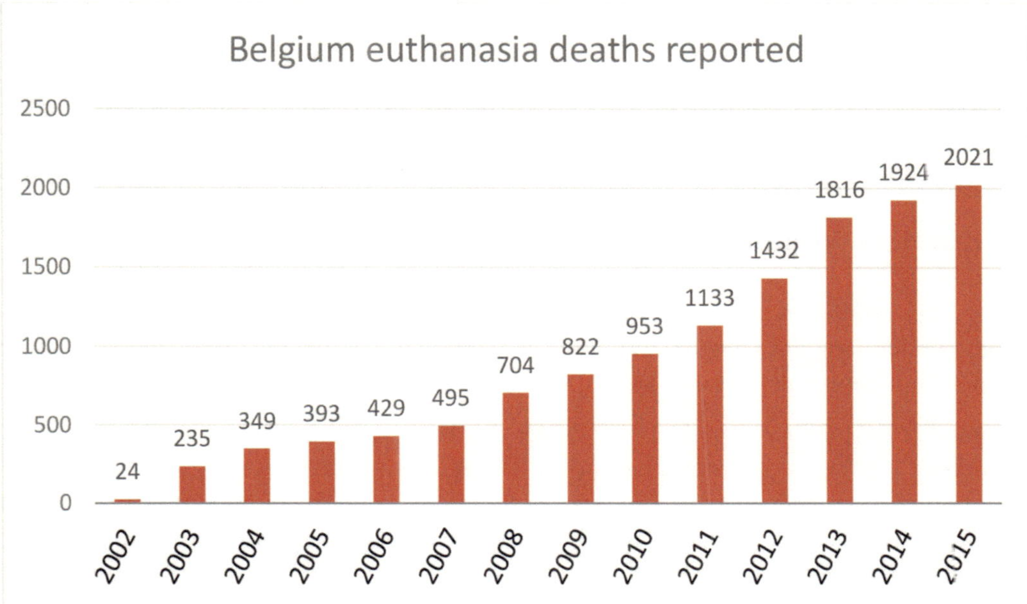

Figure 3. Deaths in Belgium reported as due to euthanasia [12, 13].

occurring. The contrast illustrates that the law is not being abused in England and Wales and is working as it should.

The numbers of such deaths vary depending on what exactly has been legalised, what the criteria are against which requests for lethal drugs are judged, and how requests are handled. But overall the numbers in each jurisdiction that has legalised some form of 'assisted dying' have been seen to increase overall year on year and have not reached a plateau anywhere (**Figures 1–3**).

3. What is 'assisted dying'?

'Assisted dying' is a euphemism that usually is interpreted as meaning physician-assisted suicide (PAS) or physician-administered euthanasia (PAE) of a person deemed to be terminally ill or who has unbearable suffering. The way that death is brought about is by administering a massive dose lethal drugs; it is unrelated to cessation of a futile treatment . Nor is it the doctrine of double effect, which is a serious adverse event from a therapeutic dose of drug, not a deliberate overdose.

For PAS, the person is prescribed a lethal dose of barbiturate, usually after preloading with an antiemetic to prevent the drug being vomited back. Such barbiturate doses are massive–about 50 times a dose that might be used in therapeutic practice–and such barbiturates themselves are very rarely used clinically today.

For PAE, the clinician, usually a doctor, injects a dose of an anaesthetic agent to induce coma. Then if the patient does not stop breathing and die rapidly, this is followed by a dose of pancuronium or a similar paralysing agent so that the patient is completely paralysed and dies of asphyxia.

Although these might seem gratuitous details, they are important to understand the process and intent in PAS and PAE, in contrast to treatment withdrawal.

4. Treatment withdrawal

When a treatment fails to achieve its therapeutic goal, or the burdens of the intervention outweigh possible benefits, it becomes futile. This situation is common in oncology when cancer escapes from the effects of chemotherapy and progresses in the face of attempted treatment.

In some circumstances, the patient may decide that enough is enough and withdraw consent to ongoing treatment, preferring instead to let nature take its course, while other patients may wish an intervention to cease and are in effect withdrawing consent for the intervention. Thus, the patient on a ventilator with advancing motor neuron disease (amyotrophic lateral sclerosis) and no possibility of improvement may decide to opt for ventilation withdrawal.

In each case, the disease process is killing the patient and their death would probably already have occurred were it not for the intervention–they are dying of their disease, not because they are being given a lethal dose of drugs to deliberately foreshorten a life that may have otherwise gone on for months or years. During their dying care must not cease, so it is completely appropriate to titrate medication as required to keep the patient comfortable and with good symptom control. But massive lethal overdoses of drugs are not used and not needed.

Another misconception is that medical science is now keeping many more people alive than previously. It is nutrition, general hygiene measures and control of epidemics through vaccination and other public health measures that have contributed far more to longevity than interventions on those who are already ill. But when illness strikes, better control of disease is certainly possible now compared to 50 years ago. Control of diabetes, statins in heart disease and thrombolysis of strokes has led to far better clinical outcomes than were previously possible. Now, the greatest societal threat to health is probably the obesity epidemic and associated chronic conditions that result in multiple comorbidities [14].

As for medical intervention keeping people alive longer, the evidence is complex. People are certainly surviving under conditions that would have killed them in the past but they are also surviving better, able to resume activities of living and for some illnesses, such as breast cancer, the disease has gone from being a death sentence to being a long-term condition, often with very long periods of remission or cure.

Improvements in general health have resulted in altered expectations in the public, fuelled by political promises in a consumerist society, which have led people to be less tolerant of debility in any form and an expectation that the healthcare system can solve the problems that are almost inevitable from lifestyle-induced disorders, ranging from tobacco and alcohol consumption, the misuse of antibiotics leading to antibiotic resistance and and obesity.

http://dx.doi.org/10.5772/65908

In general, the advances in medical science are allowing people to live much better for much longer than previously. Medical science is helping people live better and longer.

5. Criteria for assisted suicide

Different eligibility legislatures have used different criteria in their laws. It is worth examining these as there are difficulties with the verifiability of each criterion and therefore of the ability to detect if the legislation has been breached.

Importantly for patients, decisions must be based on accurate information, the patients must have the capacity to make that particular decision, and for a decision to be valid it must also be made free from coercion. Thus, the patient deciding to seek PAS/PAE must know the diagnosis is correct, the prognosis is accurate and they must be making the decision completely voluntarily.

5.1. Terminal illness diagnosis and prognosis

The majority of serious and progressing illnesses can eventually lead to death, but it is difficult to predict when death will occur in an individual. Metastatic malignant disease, with expanding deposits of malignant tumour, is relatively easy to detect and where that tumour is adjacent to a vital structure such as the spinal cord, an artery or a major airway, it is reasonable to predict that progression will result in further deterioration. Even in cancer, prediction of life expectancy – the 'how long have I got? question – is only at best be an informed guess and may be inaccurate by months or even years [15]. All too often clinicians overestimate or underestimate prognosis, leading to stories of 'they gave me three months, and here I am years later'. Although prognostic indicators in disease have been repeatedly shown to be grossly unreliable [16], Oregon's legislation requires a prognosis of 6 months, whereas Canada's legislation simply requires a doctor to state that death is likely to occur in the reasonably foreseeable future.

5.2. Mental capacity

For each decision, a person makes–and none could be greater than the decision to end your life prematurely–the mental capacity to make the decision must not be impaired. This means the person should not only be free of an illness or disorder of the mind that impairs their decision making, but also have the ability to understand the information relevant to that decision, be able to retain it and weigh it up in the decision-making process and be able to communicate this decision. In England and Wales, that is laid out in the Mental Capacity Act of 2005; there is similar legislation in Scotland and some other countries [17].

Such legislation aims to protect people from coercion and ensures that clinicians are under a duty to communicate in a way that the patients can understand. Physicians are also under a duty to do all they can to maximise the person's mental capacity, by treating reversible conditions, such as infection, and minimising the adverse effects of medications that impair capacity.

Evidence from Oregon shows that clinical depression, which leads to a particularly hopeless perspective and impairs capacity for decisions about life and death, is often not detected in assessments undertaken for PAS. Depression of itself is known to be a powerful force driving a desire for death [18], and depression and hopelessness are mutually reinforcing independent predictors of those seeking to hasten death [19]. In a small, but well conducted study one in six of those who fulfilled the assessment criteria for PAS in Oregon were found to have an undiagnosed, and therefore untreated, clinical depression; these patients were in the subgroup that then proceeded to take their own lives with lethal drugs [20].

Mental illness is frequent in seriously ill patients [21]. Suicidal thoughts have been found to occur in up to 45% of cancer patients but they usually do not persist [22]. Linked to this is the repeated clinical experience that patients react differently at different times in their illness; despair and overwhelming hopelessness can give way to hope and joy in the most unexpected circumstances [23].

5.3. No coercion

Coercive pressures are particularly difficult to detect. Clinicians do not know what goes on behind closed doors in people's homes. Coercion can be external, coming from comments that range from overheard comments through to obvious complaints about the burden the person's illness is posing on the family financially, physically and/or emotionally.

Fear of being a burden has been shown in Oregon's data over 17 years to be the second most frequently cited reason that people seek to hasten death through PAS. The perception of being a burden is itself associated with a desire for hastened death; it correlates more highly with psychological problems and existential concerns than it does with physical symptoms or difficulties [24].

Even more difficult to detect is internal coercion–the person who does not want to be a burden to the family, who is fearful of what lies ahead, perhaps who witnessed a badly managed death many years previously and is haunted by such memories, who is frightened of being undignified, confused or incontinent. Some people are unable to recognise the inherent uncertainly of life and seek to control everything around them. For them, the loss of control to a disease, which has taken over their body and is destroying their very existence, is something they cannot countenance. These people are often high achievers in life, have higher education and well-paid jobs and are used to being in command [25].

6. Clinical compassion

Campaigners cite 'compassion for the dying person' as the main driver behind demands for PAS and/or PAE, which are portrayed as the way to relieve the suffering person of their suffering. The argument has traction with the public who are fearful of pain and fearful of an existence in which they are not in control.

But amongst those who have availed themselves of lethal drugs to end their lives, Oregon's Health Department's reports show that pain comes low on the list of reasons given [10]. The main reasons given by these patients relate to existential issues, particularly being less able to engage in activities making life enjoyable (96.2%) and losing autonomy (92.4%).

There is an argument put forward that the difference between the terminally ill seeking suicide and others seeking suicide is that the terminally ill do not want to die but they recognise their impending death as inevitable and they wish to avoid suffering when dying. This would seem at first sight logical, but the difficulty is the inability to define who is truly facing death, and who, despite serious illness, can resume living well with the appropriate support. The acceptance of the inevitably of death and that disease will take its course to that death, is fundamentally different from deciding that because death seems an imminent possibility, the remainder of life, however long it is, should be dispensed with.

Some campaigners say that PAS/PAE is only intended to be available for those in the last days of life but they have not fashioned a legislative proposal that restricts requests to the period that appears to be the last days or weeks, such as the anticipated last 4 weeks of life, when prognostication stands a chance of being slightly less inaccurate. [15].

7. Autonomy

The concept of personal autonomy is also a cornerstone of arguments for legalisation of PAS/PAE. The argument is made that it is for a person and for them alone to decide the time and manner of their death. And so it may be, but that does not explain why another person (a clinician in most cases) should be involved in bringing about that death. Nor does it recognise that 'no man is an island'; we affect those around us and the very nature of society is that we are interdependent for our existence and indeed for our survival. As Onora O'Neill has pointed out, autonomy is relational [26].

It is important to recognise that the way a person dies can have profound and devastating effects on those left behind. Take the child whose mother opted for assisted suicide and who was then left feeling that his love for her was inadequate, that he had failed her by not being 'good enough' to give her a reason to live; such a sense of guilt is inconsolable and irreconcilable.

8. Dignity

Concerns about loss of dignity are also frequently given as the reason for PAS/PAE being better for a person than continued living. As Cicely Saunders said 'Dignity is having a sense of personal worth' [27] and Chochinov's work has shown that the way a person is treated by others either enhances that sense of personhood and worth, or undermines it [28].

Laws termed 'Death with Dignity Act' have an inherently misleading title as they imply that PAS of itself confers dignity on the dying person. Such an assumption is misleading. Care of the dying is not rated as highly in these countries as it is in the UK [29], where no such PAS/PAE legislation is in place and where palliative care developments have led the rest of the world in care of the dying [30]. Such perceptions that dignity in dying is synonymous with PAS/PAE mislead those other legislatures and societies considering such legislation, as well as subtly coercing patients who are wrestling with making decisions about their own lives.

9. Dying in society

As dying at home has become increasingly unusual, people have lost familiarity with dying and with death. The media, by its very nature, needs to capture viewers and listeners' attention with dramatic stories, hence the portrayal of unusual and dramatic deaths. By contrast, the thousands upon thousands of peaceful well-managed deaths that occur year in year out do not make headlines. The apparently well-managed death of David Bowie was an exception, because he was such a well-known figure, so his death was spoken about widely and opened conversations on death and dying.

The numbers of people who have not made any preparations for their own death, such as making a will, may be a reflection of fear of the unknown and a sense of denial about the reality of their own mortality. In the UK a charity, Dying Matters, has been promoting open conversations across society; the majority of those engaging in the conversation are people who are recently bereaved or had an exceptional experience of care–either good or bad–and want to talk. This lack of familiarity with dying has led to increased searching for quick solutions to complex problems, rather than a recognition that we all live with uncertainty all the time and that there are often unexpected moments of great value and tenderness as life draws to a close, if dying is planned for, accepted, and managed well.

10. Pain

Societal attitudes to death and how it should be managed have shifted over recent years, with the development of the hospice movement emanating from the UK and now adopted to greater or lesser extends around the globe. Despite advances in the science of end of life care, long-perpetuated myths about opioids have meant that these essential pain-relieving drugs remain unavailable to about 80% of the world's population. Even in those countries with legislation that enables good analgesic prescribing, misconceptions about how to prescribe such analgesics safely have led to many patients receiving inadequate analgesia to fulfil their needs. This is then witnessed by relatives, who are traumatised at seeing the person they love in ongoing pain.

The solution to the problem is to rapidly improve analgesic use and educate professionals about what pain management. As Robert Twycross has said 'you do not need to kill the patient to kill the pain' [31]. But in a search for a solution those who are unaware of what can and should be done have resorted to feeling that pain in dying patients is an unsurmountable problem to which the solution is to end the life of the sufferer.

http://dx.doi.org/10.5772/65908

11. Society normalisation

In those legislatures that have PAS and/or PAE, marked changes can be seen in attitudes to death and dying, with an inexorable increase in the numbers of premature deaths through the ingestion or administration of prescribed lethal drugs. In effect, this method of death has become normalised rather than being an exceptional event. The numbers have increased year on year, as can be seen with an 80% increase over the last 2 years in Oregon, with an eightfold increase in numbers since the 'Death with Dignity Act' came into force (see **Figure 1**). Although the absolute numbers seem small, 132 PAS deaths reported in 2015, Oregon has a very small population of only 3.8 million, which is less than half that of London.

But other changes have emerged associated with this legal change. The economic pressures of healthcare around the globe have impinged, as almost inevitably they will, on the way such decisions are viewed and reports have emerged of subtle coercive effects, whereby the costs of treatments are not funded but the far lower cost of PAS is covered by health insurance [32].

The effect on the clinical relationship between doctor and patients needs recognising too. In Oregon, there is evidence of doctor shopping, with one physician writing 27 prescriptions for lethal drugs last year even though the majority of doctors wrote none. This change reflects the process whereby patients whose doctors do not think they should have PAS are then being steered towards and assessed by physicians who do not know them as patients beyond their case notes and who have shown themselves to be more willing to prescribe lethal drugs than others.

In the Netherlands, numbers of PAE/PAS deaths are far higher than in Oregon and most are by PAE, where last year a death rate from PAE/PAS of 1 in 26 of all deaths (all causes) was reported. This may reflect several factors. Firstly, the law does not restrict the criteria to those deemed to be terminally ill, but includes those deemed to be those with intractable suffering, whatever the cause. Thus, the assessment is the patients' description of their suffering and its management, rather than also requiring consideration of the nature of the underlying pathological process. It may also reflect that the passive nature of holding out an arm for the doctor to inject lethal drugs is emotionally easier than the active gesture of raising a glass of lethal drugs and drinking the solution down, or it may reflect a different societal approach.

In Belgium, where PAE alone is legalised, the death rate appears even higher, but the statistics are harder to verify. Cases widely reported from Belgium include those who could never be classified as terminally ill, including the victim of a botched sex change operation, a prisoner with depression, twins with progressive visual deterioration towards blindness and several patients with long-term psychiatric conditions [33].

For all these cases, the term 'assisted dying' is deeply misleading as they were not dying prior to the lethal drugs being given. As such, the morality of the term itself warrants exploring. In these circumstances, terms such as 'ending life' or 'killing' would be more accurate descriptors.

12. Conclusion

Parliaments everywhere are faced with some key questions when changing the law on anything: first, does the law need changing? And second, would whatever legislation replaced its be safer overall for the whole population? better overall?

To answer the first question, it is necessary to ask what the problem is that the law is trying to solve. Some have argued that palliative care is not a universal panacea and indeed it never would be because no treatment or condition management in clinical practice ever has 100% success rate; there will always be some people for whom such approaches to their care are inadequate. But then the question of unintended consequences for the majority also needs to be explored.

At the population level, when Parliaments change the law they need clear facts on which to base their planned legislation, rather than be driven by pressure and emotive spin from campaign groups. To answer the second question, legislators need to look at the effect on the whole tenor of care in society for its vulnerable, the tensions between the costs of health and social care, and the duty to provide such care.

Legislators need to consider possible unintended consequences of legislation around PAS/PA and should look particularly at the trends from those places that have brought this into clinical practice.

Similarly, at the individual level, when patients make decisions over the options facing them, they need clear facts including information about the uncertainty around diagnosis, prognosis and other options.

For healthcare professionals themselves, there is also a need to honestly review their own roles, the financial and time pressures on them and to question their fundamental duty to patients.

In this complex debate, there is an increasing need to look at whether the law does need to be changed and how such change will alter the moral landscape. To inform that process, the evidence of the effect of legal change cannot be ignored. There is also a need to question whether the terms used are honest or misleading and to explore whether there are far safer options for patients than to licence doctors to provide lethal drugs when asked to do so.

Author details

Ilora Gillian Finlay of Llandaff

Address all correspondence to: finlayi@parliament.uk

Cardiff University, Velindre Hospital NHS Trust, Cardiff, Wales, UK

House of Lords, UK Parliament, London, United Kingdom

References

[1] https://public.health.oregon.gov/ProviderPartnerResources/EvaluationResearch/DeathwithDignityAct/Documents/statute.pdf

[2] https://leginfo.legislature.ca.gov/faces/billNavClient.xhtml?bill_id=201520160SB128

[3] http://laws-lois.justice.gc.ca/eng/AnnualStatutes/2016_3/FullText.html

[4] Termination of life on Request and Assisted Suicide Act 2002. http://www.eutanasia.ws/documentos/Leyes/Internacional/Holanda%20Ley%202002.pdf

[5] Suicide Act 1991. http://www.legislation.gov.uk/ukpga/Eliz2/9-10/60

[6] Websites which encourage suicide. Early Day Motion 256. http://www.parliament.uk/edm/2010-12/256

[7] Coroners and Justice Act. 2009, section 59.

[8] Policy for Prosecutors in Respect of Cases of Encouraging or Assisting Suicide. 2014. https://www.cps.gov.uk/publications/prosecution/assisted_suicide_policy.html

[9] Latest Assisted Suicide Figures. August 2016. https://www.cps.gov.uk/publications/prosecution/assisted_suicide.html

[10] Death with Dignity Act Annual Report. Year 18. Oregon Health Department; 2016.

[11] http://www.euthanasiecommissie.nl/actueel/nieuws/2016/april/26/jaarverslag-2015-gepubliceerd

[12] Commission Fédérale de Contrôle et d'Evaluation de l'Euthanasie. Sixième Rapport aux Chambres Législatives 2014. Chambre des représentants et sénat de Belgique.

[13] http://www.news24.com/World/News/belgian-euthanasia-cases-hit-record-high-20160127

[14] Prentice AM, The Emerging epidemic of obesity in developing countries. Int J Epidemiol. 2006;35:93-99.

[15] Higginson IJ, Costantini M. Accuracy of prognosis estimates by four palliative care teams: a prospective cohort study. BMC Palliat Care. 2002;1(1):1.

[16] Mackillop WJ, Quirt CF. Measuring the accuracy of prognostic judgments in oncology. J Clin Epidemiol. 1997;50(1):21-9.

[17] Mental Capacity Act. 2005. http://www.legislation.gov.uk/ukpga/2005/9/contents

[18] Wilson KG et al. Desire for euthanasia or physician-assisted suicide in palliative cancer care. Health Psychol. 2007;26(3):314-23.

[19] Rodin G et al. Pathways to distress: the multiple determinants of depression, hopelessness, and the desire for hastened death in metastatic cancer patients. Soc Sci Med. 2009 Feb;68(3):562-9. doi: 10.1016/j.socscimed.2008.10.037.

[20] Ganzini L, Goy ER, Dobscha SK. Prevalence of depression and anxiety in terminally ill patients pursuing aid in dying from physicians, BMJ 2008; 337: al682.

[21] Wilson KG, Dalgleish TL, Chochinov HM et al. Mental disorders and the desire for death in patients receiving palliative care for cancer. BMJ Support Palliat Care. 2014 Mar 5. doi: 10.1136/bmjspcare-2013-000604.

[22] Chochinov HM, Wilson KG, Enns M et al. Desire for death in the terminally ill. Am J Psychiatry 1995; **152**(8): 1885-91

[23] Downman TH. Hope and hopelessness: theory and reality. J R Soc Med. 2008; 101(8): 428-30

[24] Wilson KG Curran D, McPherson CJ. A burden to others: a common source of distress for the terminally ill. Cogn Behav Ther. 2005;34(2):115-23.

[25] Tolle SW, Tilden VP, Drach LL, Fromme EK, Perrin NA, Hedberg K. Characteristics and proportion of dying Oregonians who personally consider physician-assisted suicide. J Clin Ethics 2004; 15(2):111-118.

[26] O'Neill O, Autonomy and Assisted Suicide, Living and Dying Well. 2010. http://www.livinganddyingwell.org.uk/sites/default/files/LDW%20-%20Report%20-%20Autonomy%20and%20Assisted%20Suicide.pdf.

[27] Saunders C. Personal communication. 1992.

[28] Chochinov HM. Dignity and the essence of medicine: the A, B, C and D of dignity conserving care. BMJ 2007; 335(7612): 184-187

[29] The 2015 Quality of Death Index: ranking palliative care across the world. The Economist intelligence unit and the Lien foundation. 2015 London

[30] The 2105 Quality of Death Index. The Economist. London; 2016.

[31] Twycross R. Personal communication. Oxford; 1992.

[32] http://dailysignal.com/2016/07/28/when-assisted-suicide-becomes-coercive/

[33] Thienpont L, Verhofstadt M, Van Loon T, et al. Euthanasia requests, procedures and outcomes for 100 Belgian patients suffering from psychiatric disorders: a retrospective, descriptive study. BMJ Open 2015;5:e007454. doi:10.1136/bmjopen-2014-007454.